Practical Motor Boat Handling, Seamanship and Piloting

A handbook containing information which every motor boatman should know. Especially prepared for the man who takes pride in handling his own boat and getting the greatest enjoyment out of cruising. Adapted for the yachtsman interested in fitting himself to be of service to his Government in time of war

By

CHARLES F. CHAPMAN, M.E.

Editor of MoToR BoatinG

Second Edition
Revised and Enlarged

Published by

MoToR BoatinG

119 W. 40TH STREET, NEW YORK

HARVARD COLLEGE
LIBRARY

CONTENTS

CHAPTER I

Navigation Laws

NAVIGATION and shipping on the various waters of the globe, whether it be of a commercial or recreational character, is governed by certain regulations much in the same way as traffic on land is regulated. Instead of coming under the jurisdiction of the various States, as is the case on land, navigation and traffic on water comes under the jurisdiction of the Federal Government, except on inland waterways which are entirely within the limits of one State.

Federal jurisdiction is divided into three principal classes, which may be referred to as the International, the Inland, and the Pilot Rules.

International Rules

The International Rules govern navigation on waters which do not come within the jurisdiction of any particular country; for example, navigation on the high seas beyond what is technically known as the three-mile limit comes under the jurisdiction of the International Rules. These rules were drawn up at a conference of a number of the maritime nations of the world held about 1890. The various nations which were represented by delegates at this conference agreed to certain uniform and standard regulations, which should govern the ships of their nations on the high seas. They adopted and agreed to abide by the regulations which have been in force since this conference.

Inland Rules

Navigation on the waterways of the United States, which are tributary to the high seas, or are not within the confines of a particular State, are governed by regulations adopted by the Congress of the United States. These regulations are known as the Inland Rules, of which there are several sets applying to inland waters in particular localities.

Pilot Rules

Congress has given power to the Board of Supervising Inspectors to enact laws governing the navigation of boats on the inland waters of the United States, which must not be contrary or opposed to the inland laws. These are known as the Pilot Rules.

The Pilot Rules vary somewhat for the different localities, and are published by the Department of Commerce in book form entitled "Pilot Rules for Certain Inland Waters of the Atlantic and Pacific Coasts, and the Gulf of Mexico," as is well known by every motor boatman.

Motor Boat Rules

Included in the book of Pilot Rules is the Motor Boat Act of 1910, which governs the rules for lights, equipment, etc., to be carried on boats of less than 65 feet in length, which are technically known as motor boats. In other words, a motor boat is any vessel operated by machinery that is less than 65 feet in length, other than tugboats propelled by steam.

The Board of Supervising Inspectors, in accordance with the power granted it by Congress, has the right to make certain regulations of a local nature arising from conditions in the particular locality to which they refer, such as speed regulations, signals for vessels requiring the assistance of police or fire boats, rules referring to the operation and mooring of dredges, marking of wrecks, etc.

War Department

The War Department is entrusted to certain regulations related in a certain way to navigation, such as the improvement of waterways and channels, the establishment of pier head lines, certain dredging operations, bridge regulations, etc. Congress has conceded certain rights to be within what is known as the police power of the respective States.

Fortunately, the International, Inland, and Pilot Rules are practically identical in their wording and intention, differing only in the minutest details, so that from the standpoint of the motor boatman there is no need to consider that there exists more than one set of rules for his guidance.

CHAPTER II

Meeting and Passing

THE major portion of the navigation laws naturally refer to boats under way. A boat is considered to be under way, according to law, when she is not at anchor, aground, or made fast to the shore.

Special Circumstances When Risk of Collision Exists

The fundamental basis of the laws governing right of way between two boats on different courses provides that one shall have the right of way and must hold her course and speed, while it shall be the duty of the other to give way or keep clear of the boat having the right of way. However, the law specifically states that in obeying and construing the rules of the road at sea, due regard must be had for all dangers of navigation and collision, and to any special circumstance which may render a departure from the rules necessary in order to avoid immediate danger. The rules also specifically state that when for any reason whatsoever a vessel which has the right of way finds herself in such a position that an accident or collision cannot be avoided alone by the boat which is supposed to give way, then the boat having the right of way must do everything within her power to prevent a collision. This rule practically means that to a greater or less degree the responsibility for a collision is placed upon both boats.

Duty in Case of Accident

Should an accident or collision occur, it is the duty of the person (or persons) in charge of each boat to stand by the other until he has ascertained that she is in no need of further assistance, and it is his duty to render to the other boat, her crew, and her passengers such assistance as may be practical and necessary in order to save them from any danger caused by the collision—so far as he can do so, that is, without danger to his own boat, crew, or passengers. He must also give to the person in charge of the other boat the name of his own craft, and the port to which she belongs.

Precautionary Measures

Certain whistle signals are provided by law, to be given by one boat in order to indicate to the other her course and action. Although not required by law, it is essential that the boats take other action than the mere giving and answering of whistle signals. It is a well-known fact that whistle signals are of little value on a motor boat where the noise from machinery or from other causes is excessive. Very often whistle signals given on the kind of whistle which exists on many motor boats to-day cannot be heard for any appreciable distance. It is far better that the course which one intends to take should be indicated by swinging the boat's bow decidedly in that direction. This action can be seen by the person in charge of the other boat, and in many cases it is decidedly more convincing than an exchange of whistle signals. It is essential that in passing another boat you give it as wide a berth as possible. Nothing has ever been gained by passing close to, and much has been lost.

When to Give Whistle Signals

Whistle signals should only be given when boats are actually in sight of one another in the day-time, or when the lights are visible at night, when danger of collision exists. Whistle signals should never be given or exchanged when these conditions do not exist. They must never be used under any circumstances in the fog or when the exact location of a vessel cannot be accurately determined.

The Helm

In the various navigation rules, as well as in many situations which one meets aboard a motor boat, the helm is often referred to. It must be remembered that the boat's helm is not her wheel or her rudder. When the order is given to port one's helm, action should be taken so that the boat's bow will swing to starboard. When a starboard helm is referred to, the boat's bow must be swung to port. In other words, the helm more nearly applies to the tiller of the old sailing days than to the wheel or the rudder. Wheels as they are rigged to-day may turn in the direction in which the boat's bow is swung, or in the opposite direction. Thus it will be seen that the term helm cannot refer to the wheel.

The Three Situations

In the case of two boats approaching each other, we find three cases called meeting head-on, overtaking, and crossing, respectively. The first two situations mentioned are clearly defined, and the third must be assumed to exist when the boats are not meeting head-on, or in an overtaking position.

Boats are assumed to be meeting head-on when thei: masts can be seen in line, or nearly so. (Nearly so, has been ruled to mean one-half a point either side of dead ahead.) They are assumed to be in an overtaking position when one boat is approaching the course of another from more than two points abaft the beam of the leading boat, or when at night the side lights if correctly placed cannot be seen. In all other situations where the courses intersect, either at right

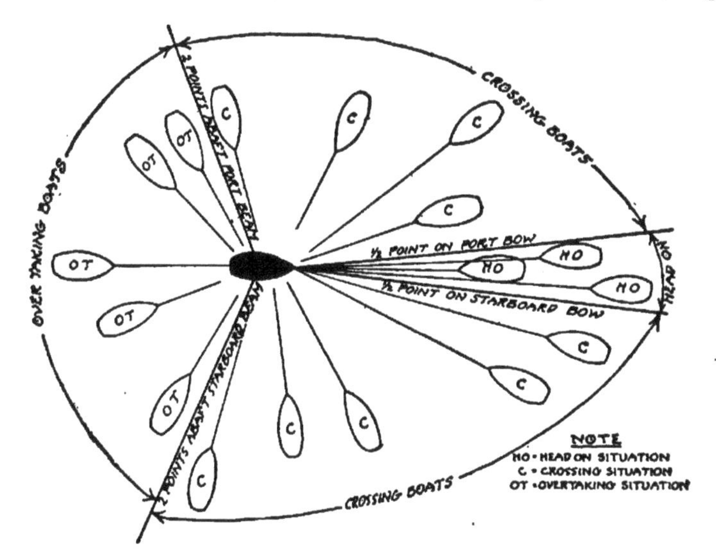

Fig. 1A. The three situations—boats meeting head-on, overtaking, and crossing. Relative to the black boat, the boats marked H O are in the head-on situation; those marked O T are overtaking; and the boats marked C are in a crossing position

angles or obliquely, they are assumed to be crossing. (See Fig. 1A.)

Meaning of Points

To get a clearer understanding of the meanings of the various terms used on shipboard as well as the meaning of head-on, overtaking and crossing, Fig. 1 may prove of value. It will be noticed that a circle circumscribed around the boat is divided into 32 equal parts, known as points, a particular name being given to each one of these divisions. Any object located on a line extended from the center of the boat and passing through its bow is said to be dead ahead.

An object on the line which is drawn on the first 1/32 of the circumference to the right is said to bear one point on the starboard bow. When it is on the second line as shown, it is said to bear two points on the starboard bow. If on the third line an object bears three points on the starboard bow. An object on the next line, which is 45 degrees or ⅛ the way around, is said to bear four points, or broad on the starboard bow.

As we continue around to the right, the next bearing will be called three points forward of the starboard beam; the next two points forward of the starboard beam; and the next one point forward of the starboard beam. An object on the starboard side on a line drawn at right angles to the fore and aft line of the boat is said to be on the starboard beam.

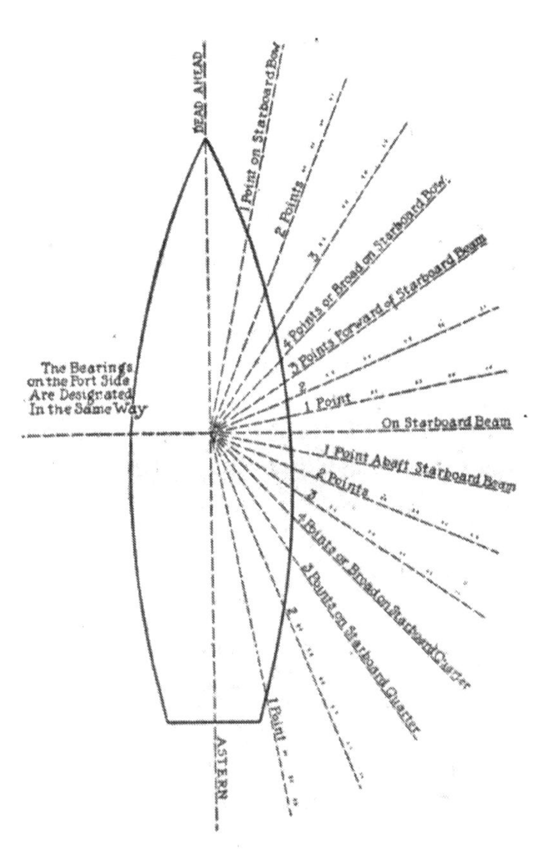

Fig. 1. Names and location of various points and bearings

In a similar way the points aft of the starboard beam are called one point abaft the starboard beam; two points abaft the starboard beam, and three points abaft the starboard beam. The next point is called four points or broad on the starboard quarter. The next, three points on the starboard quarter; two points on the starboard quarter and one point on the starboard quarter. An object on the next line, which is a line extended from the bow and passing through the center of the ship is said to be astern. The bearings on the port side are designated in a similar way.

From the diagram it will be seen that we have three principal designations used in referring to different parts of the boat, namely its *bow*, which extends from dead ahead through four points to starboard or port; the *beam* of the boat, which extends through eight points on either side; and the boat's *quarter*, which extends four points forward of astern on each side.

Right of Way Diagram

In determining which of two boats whose courses are approaching each other has the right of way, Fig. 2 will be of aid. If the diagram of the boat is assumed to be your ship,

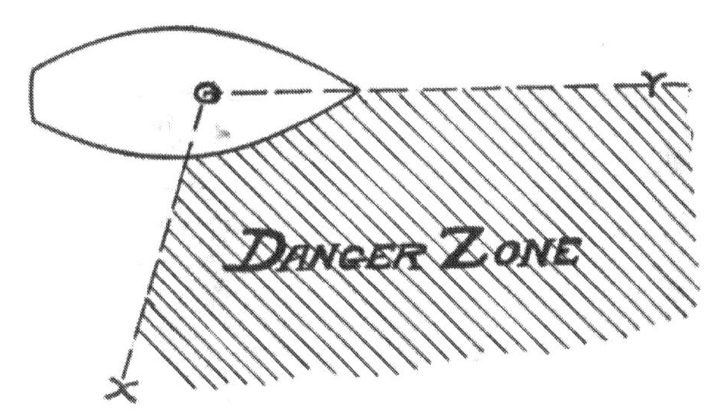

Fig. 2. Boats approaching your course within the danger zone have the right of way over your boat when danger of collision exists

then you would have the right of way over every boat approaching your course, with the exception of those boats which were approaching you in the shaded section marked "Danger Zone." Boats in the danger zone approaching your course would have the right of way over you, and it would be your duty to keep clear of such boats. In all other cases where boats were approaching your course in the clear section, your craft would have the right of way, and the obligation to keep clear would fall upon the other boat. In other words, the danger zone extends from dead ahead, represented by the line *OY*, around to two points abaft the starboard

Fig 3. Boats meeting head on. Proper action—each boat blows one blast of her whistle, ports her helm, and they pass port side to port side

Fig. 4. Boats crossing. The boat on the reader's left has the right of way, and after giving one blast on her whistle she should maintain her course and speed. The other boat must pass astern

beam, which is represented by the line *OX*. At night, boats approaching your course within the danger zone would show you their red light, another symbol of danger, and an indication that you must give way. At night, boats approaching your course anywhere in the clear section would show you their green light, a signal for you to hold your course and speed.

Proper Location for Steering Wheel

Fig. 2 should bring home to you why it is preferable to have one's steering wheel located on the starboard side of the boat rather than on the port side, if one of the two locations must be chosen. One will realize that it is essential that the helmsman should have an unobstructed view of the danger zone, as it is boats within this zone which give him the chief concern. To get the clearest view of the danger zone, the steering wheel should naturally be located on the starboard side. One does not have as much concern as to what is happening on the port side, as boats approaching your course on that side must give way to you.

Meeting Head-On

Of the various meeting and passing situations of two boats, meeting head-on (as shown in Fig. 3) is the simplest one. In such a case it is the duty of each simply to turn to the right exactly as two vehicles on land would do. As an indication that this action is to be taken, one boat should sound one blast on her whistle, swinging her bow at the same moment. When this action is understood by the other boat, she should reply by one blast of the whistle, also swinging her bow simultaneously. The boats will then pass port port side to port side.

Courses Parallel

When two boats are on parallel courses, but each course is so far to the starboard of the other that no change of course is necessary in order to allow the boats to clear, two blasts of the whistle should then be given by one boat, which should be acknowledged by two blasts from the other boat, each holding her course and speed, and they pass clear of each other starboard side to starboard side. This is the only situation where it is allowable to use two whistles in passing. All other cases where two whistles are used are illegal, and should be avoided.

Overtaking

A boat is considered to be overtaking another when she is approaching the course of the leading boat from more than two points abaft the beam of the leading boat. In such a situation the rights all rest with the leading boat, the overtaking vessel having no rights whatsoever. In a situation of this kind if the overtaking boat desires to pass on the starboard side of the leading boat, she may ask permission to do so by giving one blast on her whistle. If the leading boat believes it is safe and practicable to allow the overtaking boat to pass on her starboard side, she will answer by one blast of her whistle, in which case the overtaking boat will pass to starboard, being careful not to interfere with the course or rights of the leading boat. However, should the leading boat, for any good reason, believe that it is undesirable to allow the overtaking boat to pass her on her starboard side, she may refuse permission by giving the danger signal, four or more short blasts on her whistle, in which case the overtaking boat has no alternative other than to stay astern. In the course of a short time the overtaking boat may again ask permission to pass the leading boat on the starboard side by giving one blast of her whistle, and the leading boat again has the right to exercise the prerogative that has just been explained.

Should the overtaking boat desire to pass the leading boat to port, her signal would be two blasts on the whistle, which would be answered by two blasts by the leading boat, if she considered that conditions warranted the action requested by the overtaking boat; otherwise she would sound the danger signal, and proceed exactly as before.

When doubt exists in the minds of the master of one boat as to whether his craft is an overtaking boat or a crossing boat, he must assume that he is an overtaking boat, and be governed accordingly.

Crossing Courses

Courses which are crossing, or which may be said to be meeting obliquely, probably form the most common situation. From Fig. 2, which illustrated the danger zone, it will be remembered that the boat, which has the other on her port bow is considered to have the right of way. In such a situation it is the privilege and duty of the boat having the right of way to maintain her course and speed. The boat not having the right of way must give way in every instance.

The proper signal for the boat having the right of way

when two courses are meeting obliquely is one blast of the whistle. (See Fig. 4.) This should be answered immediately by one blast from the other boat. The boat not having the right of way must then pass astern of the boat which has the right of way. If necessary, the vessel having to give way must slow down, stop, or change her course in order to allow the other boat to pass ahead of her.

Assuming that no whistle signals have been previously given, if for any reason the boat not having the right of way desires to, she may ask permission to pass ahead of the right-of-way boat by giving two blasts on the whistle. If the right-of-way boat is so inclined, she may grant this permission by answering with two short blasts of her whistle. However, in granting this permission by giving two blasts of the whistle, it is understood by the other that she may pass ahead at her own risk. Such a reply does not of itself change or modify the statutory obligation of the giving-way boat to keep out of the way as before, nor does it guarantee the success of the means she has adopted to do so. In other words, should an accident occur, the responsibility will rest entirely with the boat which has not the right of way, even though the fault seems to lie entirely with the other craft. This is a situation which is very common on the waterways of our country. But motor craft should always be careful to avoid it as it is entirely illegal.

Should the boat not having the right of way request permission to pass ahead of the other boat by giving two blasts of her whistle, and should the right-of-way boat not desire to grant this request or permission, she will sound the danger signal, in which case both vessels must stop, and be absolutely sure of the action of each other before proceeding.

Boats Backing

In the instance of one or more boats backing, the case is comparatively simple, as the stern of the backing boat for the time being is considered her bow. Passing signals are exchanged exactly as though the boats were proceeding ahead, considering for the time being that the boat's stern is her bow.

Boats Coming Out of a Slip

When boats are backing out of slips, or away from wharfs or piers, the rules of the road do not apply until the boat is entirely clear of the slip. In other words, she has absolutely no rights of way until she is clear. As a boat starts backing out from such a landing she is supposed to give one long blast on her whistle. As soon as she is clear of such obstruc-

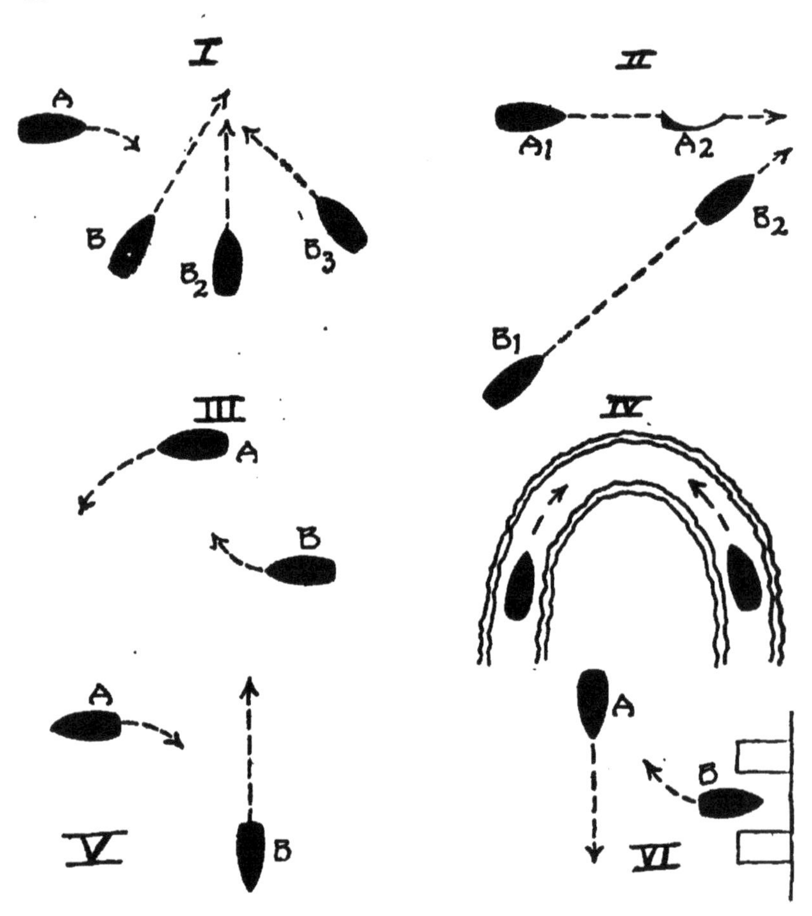

Fig. 5. I, Crossing boats; II, Overtaking boats; III, Boats on parallel courses; IV, Meeting in a winding channel; V, Backing; VI, Boat coming out of slip

tions the regular rules of the road and rights of way apply. However, no craft has the right to run so close to a pier line that entrance or exit from slips and wharves will be blocked. Ferry boats and others must be given a reasonable amount of space for maneuvering purposes at the entrance to their landings.

Boats on Parallel Courses

We now come to the situation of two boats on parallel courses where one desires to cross ahead of the other. As the two boats are on parallel courses, neither has the right of way over the other, strictly speaking, but in many instances one boat desires for good reasons to change her course, and pass across ahead of the other boat. For example, *A* in Fig. 5 III being the faster boat, desires to change her course, as indicated by the arrow, and pass across *B*'s bow. The question is whether she should give one whistle to indicate this course, or whether two blasts will be necessary.

There is considerable difference of opinion in the minds of motor boatmen as to the proper signal, as many will hold that this is a two-whistle situation—but such is not the case. If *A* desires to cross *B*'s bow, she shoul' give one blast of her whistle, and this should be answered by one blast from *B*'s whistle, upon which *B* should pass astern of *A*. Probably the confusion in a situation of this kind results from the fact that many have learned the old rule that one whistle means "I am directing my course to starboard," and two whistles is an indication that "I am directing my course to port." In the case just cited, this rule does not hold good, and the sooner one can eliminate it from his memory the better. However, there is a rule which is much easier to remember than the old one, and it holds good in every case.

As everyone knows, there are two terms used in expressing direction of a boat, namely, port and starboard. The former word has one syllable, and the latter two syllables. We also have a signal of one whistle, and a signal of two whistles. If one keeps in mind that one whistle always refers to the word of one syllable—namely, port—and two whistles always refers to the word of two syllables—namely, starboard—he will have no difficulty in any of the passing situations. Wherefore, one whistle always means "I am going to pass you on my port side," and similarly, two whistles means "I am going to pass you on my starboard side."

One whistle, one syllable, port—two whistles, two syllables, starboard.

Overtaking and Crossing

In the upper left hand part of Fig. 5, the boat *B* in any of the positions marked *B*1, *B*2 or *B*3 would have the right of way over *A*, and it would be the duty of *A* to keep clear. However, in the upper right hand diagram the boat *A* would have the right of way over the boat *B*, for the simple reason that in the first case boat *B* is a crossing boat, and in the second instance an overtaking boat.

Meeting in Winding Channel

When two boats are approaching each other in a winding channel (See IV, Fig. 5), they must be considered as meeting head-on, and not as meeting obliquely. In such a case neither has the right of way, but it is the duty of each to swing to starboard after one blast of the whistle, and pass port side to port side.

Rights of a Sailing Vessel

A sailing craft (See Fig. 6) has the right of way over a motor-driven craft in every instance except one. The one exception is when the sailing craft is in such a position as to be considered an overtaking boat. In such a case the motor boat or leading boat would have the right of way over the sailing vessel. In all other situations the motor craft must give way to the sailing vessel.

Auxiliaries

Boats of the type known as auxiliaries, capable of being operated under both sail and power, or under either alone,

Fig. 6. Sailing craft have the right of way over power vessels, except when the former is overtaking the power vessel from more than two points abaft her beam

are subject to certain rules of the steamboat inspectors. In some cases inspectors have ruled that an auxiliary when proceeding under sail and power would have the rights of a sailing craft; in others that she would have only those of a motor craft. However, at night a boat proceeding under both sail and power is required to carry only the lights of a sailing vessel. An auxiliary, operating under sail alone, of course, has the rights of a sailing vessel, and when operating under power alone is classed as a power vessel.

Cross Signals

Whistle signals given by one boat must always be answered by a similar whistle from the other vessel. That is, one whistle must always be answered by one, and two whistles answered by two. If for any reason the signal cannot be answered by a similar whistle, it must be answered by the danger signal, which is four or more short blasts of the whistle. It must never be answered by a cross signal; that is, two whistles must never be used to reply to one, nor one to reply to two. When the danger signal is given by either of two boats it is the duty of each to proceed only with the greatest caution, or stop and reverse if necessary. In other words, the danger signal is a signal that the action of one vessel is misunderstood by the other, and neither boat should proceed until the proper signals have been given, exchanged, and understood.

CHAPTER III

Lights for all Classes of Boats

IN discussing the subject of the proper lights for the various classes of vessels we should keep in mind the fundamental laws mentioned in the first chapter. Although on first thought it might appear that with three or four sets of laws (the International, the Inland, and the Pilot Rules) to guide us, confusion would result in many instances, as the laws in themselves might conflict, we find upon analyzing the situation that this is not the case, as in only a few instances are the rules themselves conflicting, and then in only minor respects. Therefore, for the purpose of this general discussion, we can very readily assume that only one set of laws governs the proper lights to carry.

There are many types and kinds of vessels, and it is but natural that the laws should provide lights so different in their characteristics that different types of vessels could be readily distinguished. In this respect the laws in most instances are strikingly efficient, yet they have not been made so complex as to become cumbersome.

Types of Craft

The principal division into which types of floating craft are divided are sea-going vessels, inland vessels, tow boats, sailing craft, ferry boats, barges and canal boats, scows, rafts, etc., and last but not least, motor boats. This subdivision of motor boats is further divided into three classes, depending upon the overall length of the motor boat.

Fundamental Rules

In familiarizing oneself with the lights one has to remember very little. The only colors used for lights are white, red and green, and these are arranged to show in only four different ways—namely, for 10, 12, 20, or 32 points around the horizon, the last, of course, being the light which may be seen from all directions. (See Fig. 7.) Red and green lights each show for 10 points, irrespective of the type of vessel on which they may be used. White lights showing from ahead are invariably arranged to be visible for 20 or 32 points. Lights arranged to show from astern are invariably visible for 12 or 32 points. There are no other combinations than those just mentioned, so if one gets this fundamental fixed in his mind he will have no trouble in remembering how the various lights are arranged to show.

Lights for Motor Boats

As the lights to be carried on motor boats are of the greatest interest to us, we shall consider that phase of the subject first. As it is not practicable to photograph boats at night for the purpose of showing their lights, a system has been adopted which will better illustrate the points that are to be brought out. Fig. 8 makes clear the symbols that have been adopted for the two classes of white, the green, and the red lights, respectively.

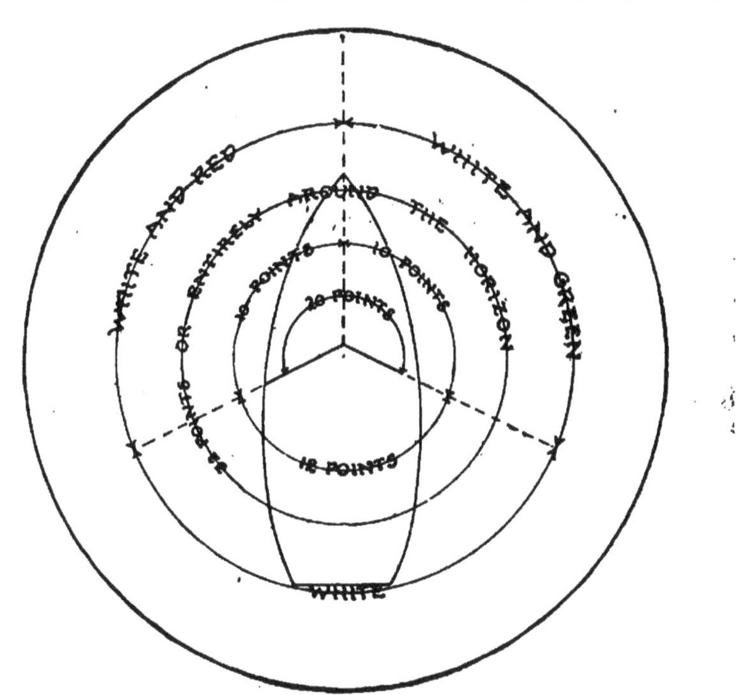

Fig. 7. Diagram showing range of visibility of lights on all types of boats

As was mentioned above, motor boats for the purpose of lighting are divided into three classes, according to their overall length. These are, Class 1, consisting of boats under 26 feet in length; Class 2, boats measuring 26 to 40 feet in length; and Class 3, boats of 40 to 65 feet in length. In only a few respects does the lighting system vary between the different classes, and so it is possible to consider motor boats as a whole, simply mentioning when the system of lighting varies from one class to another.

The boat shown in Fig. 8 probably falls in Class 3; that is, she has a length of somewhere between 40 and 65 feet. It will be noticed that on the forward part of her awning is placed a white light showing ahead, and directly under this her green starboard light, which shows from directly ahead to two points abaft the starboard beam, or a total of 10 points. Although this arrangement may be legal, it would be better to have the white light much nearer the bow and on a level with the green starboard light. The symbol used indicates that the white light at the stern of the boat shows completely around the horizon. This light is placed higher than the forward white light, which is proper and necessary. In other words, a white light aft show-

Fig. 8. Lights carried by a motor boat under way

ing completely around the horizon must be higher than the forward light showing directly ahead. The reason for this will be brought out later. Too many motor boat owners make the mistake of hanging this after light below the awning, where it cannot be seen completely around the horizon, and others err in not placing it higher than the forward light. Obviously, there is a red light on the port side showing for 10 points, placed similarly to the green starboard light.

Class 2 motor boats are lighted exactly like Class 3, the only difference being in the sizes of the lenses of the lights required by law, and the fore and aft length of the screen used to shield the red and green side lights. The law requires that the lenses of all lights shall be of fresnel or fluted glass.

Class 1, comprising motor boats of less than 26 feet length, may be lighted in the same way, or, if their owners prefer, they may substitute a combination red and green light in the bow in place of the three forward lights of Class 2 and 3. In such a case the starboard light must be so constructed as to show a green light from directly ahead to two points abaft the beam on the starboard side, and the port lens to show a red light over 10 points on the port side.*

Motor Boats Under Sail and Power

Motor boats under sail and power carry only the red and green side lights, each properly screened to show over 10 points

Fig. 9. A motor boat under sail and power carries only the colored side lights

of the horizon. (See Fig. 9.) Motor boats under sail and power never show a white light except upon being overtaken by another vessel, when a white light is temporarily shown over the stern of the auxiliary.

Lights for Inland Steamers

Inland vessels of the type shown in Fig. 10 are lighted exactly like boats in Classes 2 and 3—that is, a white light forward

*For sizes of lenses and other details, see chapter on Equipment Required by Law (page 51).

Fig. 10. Lights carried by an inland steamer

showing over 20 points, red and green side lights each showing over 10 points, and a range light aft showing completely around the horizon. The white light, as previously mentioned, should be placed as near the bow as possible; the side lights placed a little farther aft on the same level, and the range light placed higher than the bow light.

Sea-Going Vessels

Sea-going vessels differ only slightly in their lighting particulars. (See Fig. 11.) The forward light, as usual, shows over 20 points, but is generally placed about halfway up on the foremast. The customary red and green lights are carried, but on sea-going vessels they must naturally be placed somewhat lower than the forward light. The only point of difference in the lighting of sea-going and other vessels consists in the arrangement of the after range light, which generally shows ahead for 20 points instead of completely around the horizon. This should bring to mind the situation that viewing a sea-going

Fig. 11. Lights carried by a sea-going vessel

Fig. 12. Steam yachts are generally lighted in the same manner as sea-going vessels

vessel from astern there would be no light visible. This is perfectly true, but the law provides that on any type of similar vessel where there is no light visible from astern, an additional light may be carried, generally low down, visible from astern only, or around 12 points of the horizon. It is a fact that most sea-going vessels carry this light on their taffrail.

Fig. 13. Sailing vessels under way carry red and green side lights

Steam Yachts

Yachts which might be termed sea-going are lighted in exactly the same way as sea-going vessels (See Fig. 12); that is a white light on the foremast showing ahead for 20 points, a range light on the mainmast showing ahead for 20 points, and placed higher than the light on the foremast, and the usual red and green side lights, each showing 10 points.

Lights for Sailing Vessels

Sailing vessels (See Fig. 13) carry the red and green side lights, each showing for 10 points, and no other lights, except upon the approach of a vessel from astern which is overtaking the sailing vessel from such a position that a side light is not visible, when the flare-up or other white light is shown over the stern to attract the attention of the approaching vessel.

Ferry Boats

Ferry boats carry two central range lights showing completely around the horizon, placed at equal altitudes forward and

Fig. 14. Ferry boats carry two central range lights at equal altitudes above the water showing all the way around the horizon, the customary colored side lights, and a special distinguishing light placed above the central range lights

aft, and generally on top of the pilot house. (See Fig. 14.) In addition, the usual side lights are carried. A ferry boat may carry an additional light showing completely around the horizon, and usually hoisted on one of the side flag staffs about 15 feet above the white lights. This light is used to distinguish the particular line to which the ferry boat belongs, and different colored lights are used for the different lines. For example, in New York harbor, this light on the boats on the Pennsylvania Railroad is red, that of the Erie Railroad white, and that of the Lackawanna green. Ferry boats which are not of the double-end variety carry the usual white lights, and colored side lights required by law to be carried by steam vessels navigating those waters.

Harbor Tugs

In the matter of lights for tow boats, we find two principal classes, which might be called sea-going tow boats, and harbor tugs, although the lights for harbor tugs are restricted to such places as New York Harbor, Long Island Sound, Hudson River, and adjacent waters. Harbor tugs (See Fig. 15), with a tow, on these waters, carry the usual red and green side lights, and in addition either two or three white lights vertically arranged, showing completely around the horizon. Whether they carry two or three of these lights depends upon the length of their

Fig. 15. The lights carried by a harbor tug towing one vessel only, or when the length of the tow is less than 600 feet if more than one vessel is towed. The vessel being towed carries the red and green side lights only

tow, provided more than one vessel is being towed. When the length of the tow of two or more vessels measures less than 600 feet, then two lights are carried. If this length exceeds 600 feet, then three lights vertically arranged are used. A vessel being towed carries only the red and green side lights, without the white lights. It should be noted that the term "Vessel being towed" does not include types of boats which fall into classes usually known as barges, canal boats, scows, rafts, etc. These latter types have particular lightings, which will be explained shortly. When only one vessel is towed, the tow boat shows two white lights, irrespective of the length of the tow.

Ocean-Going Tow Boats

An ocean-going tow boat (See Fig. 16) displays the two or three white lights, according to the length of the tow, but usually

Fig. 16. Lights carried by a sea-going tug

the white lights instead of being placed so that they show completely around the horizon, are arranged on the forward part of the foremast, to show ahead for 20 points only, and in a vertical line. The usual red and green side lights are carried, and in addition to this, a white light, showing astern only, may be carried. This latter light is the only one visible from astern, and is used by the boats towed to steer by. It is interesting to note that the difference regarding range lights on inland and sea-going steamers applies in a similar way to harbor and sea-going tugs; in other words, on inland steamers, the range light shows completely around the horizon, while on sea-going steamers it shows ahead only. Similarly on harbor tugs, the range lights show completely around the horizon, while on sea-going tugs they show ahead only.

Barges and Canal Boats

Barges and canal boats towed on certain inland waters, as for example, the Hudson River, New York Harbor, Long Island Sound, etc., have special lightings. When they are towed in tandem there is a white light placed on the forward and after ends of each barge, with the exception of the after end of the last boat in the tow, which, instead of showing one white light,

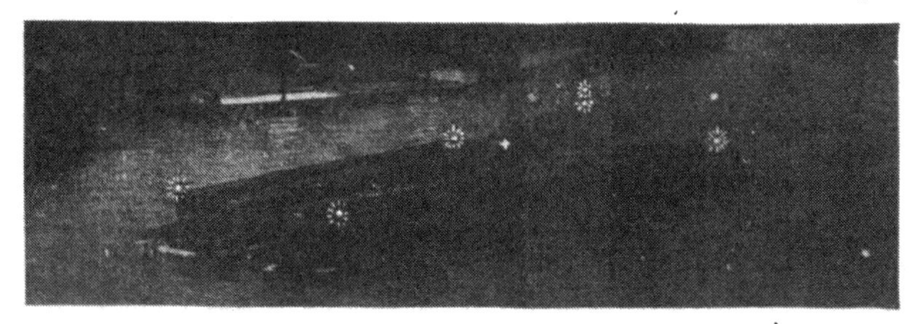

Fig. 17. Lights carried by railroad floats when being towed

displays two white lights horizontally arranged. All of these white lights are placed to show completely around the horizon.

Car Floats

Fig. 17 is intended to show the lighting of railroad barges or scows towed alongside with a tug between them. It will be observed that a white light showing completely around the horizon is placed on the two outer corners of the two barges. Had there been only one railroad float alongside the tug, then only white lights on the two outer corners would have been used. It will also be noted that the tug is carrying two white lights vertically arranged, showing that the length of tow is less than 600 feet. Furthermore, the tug is carrying the usual red port light. Had the height of the float been sufficient to hide this port light from its proper view, the port light would have been carried on the outer edge of the port barge or float. The same is true, of course, of the green light.

Pilot Vessels

Pilot vessels (See Fig. 18) on their stations carry in addition to the red and green side lights two other lights on their main mast, showing completely around the horizon, the uppermost light of the two being white, and the lower one red. Pilot vessels, while not engaged on their station on pilotage duty, carry similar lights to those of other vessels. When a pilot vessel is engaged on a station on pilotage duty, and is at anchor, she does not carry the red and green side lights, but continues to display the white and red mast headlights.

Fishing Boats

Fishing vessels of more than 10 tons when under way, but not having nets or lines out in the water, show the same lights

Fig. 18. Lights carried by a steam pilot vessel on her station

as other vessels. However, when such vessels are engaged in trawling, dredging or fishing, they exhibit from some part of the vessel, where they can best be seen, two lights. One of these lights is red, and the other white, and the red light is above the white.

Fig. 19 shows diagrammatically the arrangements for lighting the various classes of boats when towed, including the lights of ocean-going barges when being towed in tandem. These barges carry red and green side lights, and in addition a white light at the stern of each barge, showing astern only, with the exception of the last barge in the tow, which instead of carrying one white light aft showing astern only, carries two lights, horizontally arranged, showing completely around the horizon.

Ocean-going barges when towed alongside, if their height is sufficient to obscure the side lights of the towing vessel, carry

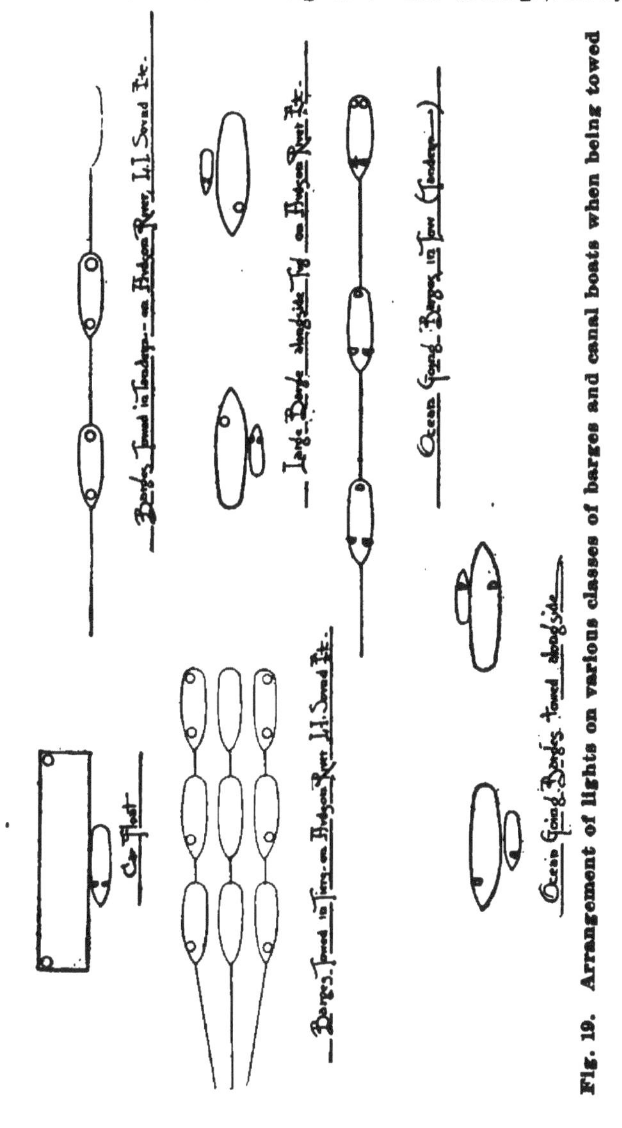

Fig. 19. Arrangement of lights on various classes of barges and canal boats when being towed

the red or green starboard light on the proper side, to take the place of the obscured light.

Value of Range Light

Figs. 20-21 illustrate the value of the after range light. Fig. 20 shows the various ways which a boat might be heading when both the red and green side lights are visible. Consider for a moment that your own boat is the one lettered *A*, and the side lights of boat *B* can be seen ahead of you. If boat *B* had no after range light it would be impossible for you on boat *A* to determine the course of *B* with any degree of accuracy. Boat *B* might be heading as shown by *B*1, *B*2 or *B*3, or, in fact, any position between *B*1 and *B*3, and from your position on any of the boats lettered *A* you could not determine how *B* was heading. You might assume that the two boats were heading as indicated by *B*1 and *A* directly under it. In such a case you would sheer off to starboard in order to pass port side to port side. As a matter of fact, while you assumed the boats to be in the position of *B*1 they might in reality be heading as *B*3 is indicated, and when you threw the head of your boat *A* across the path of *B*3, a collision would be inevitable—there would be no way by which you could determine what the direction of *B* really was. This is not an exceptional case, but a common one,

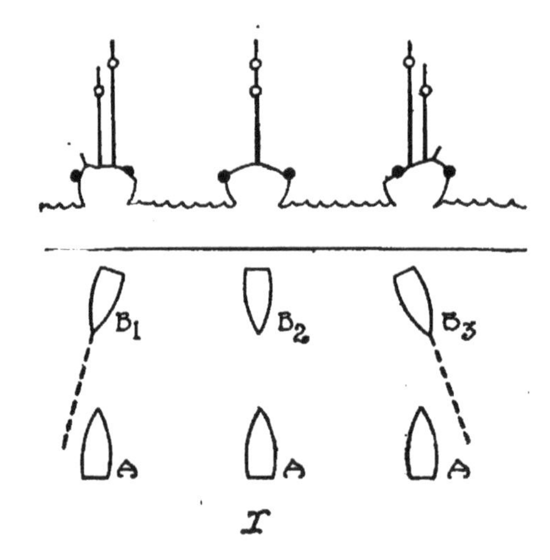

Fig. 20. Various ways a boat might be heading when both side lights are visible

and one which every boat must be in before passing clear of another boat when they are meeting head-on.

The after range light, if properly placed, is a key to the whole situation, as will be observed by the boats *B* in the upper part of the diagram, which are in the position indicated by *B*

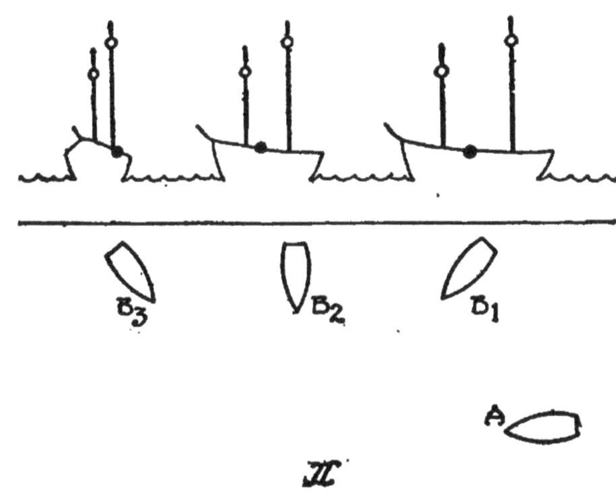

Fig. 21. When only the red side light is seen a very dangerous situation might develop. It is the range light which indicates the correct heading

directly under them. The relative position of the forward and after range lights immediately determines the exact positions of *B*. Doubt no longer exists as to the exact course of *B*, or what action *A* should take to properly clear *B*. When the range lights are directly over each other it is clear that a vessel is approaching you dead head-on, but when her course is changed even in the slightest, the range lights will open out, the lower one drawing away from the upper in the direction in which the boat's bow is changing. Without the range lights the boat's course might change several points before being made evident from the side lights. It is even possible that the course of the approaching vessel is away from the course of one's own boat when the side lights are first sighted, and that she swings around toward your course without this being detected from the side lights as the boats draw closer together. It is a very dangerous position if the range lights are not properly placed, and one which requires great caution.

A situation not unlike the last when both side lights were to be seen develops when only one of the side lights is visible. In Fig. 21 it will be noted that the red light is to be seen, while the green starboard light is hidden. Without a range light it would be impossible for you to determine from your position on boat *A* whether boat *B* was heading as indicated by *B*1, *B*2 or *B*3. You might plan your action assuming that the approaching boat was heading as indicated by *B*1, when in reality she might be headed as indicated by *B*2, or even *B*3, and yet it would be impossible for you to determine the exact heading. Again the after range light solves the problem. From the upper part of this diagram the boats are arranged to show exactly as they would from your position on *A*. Looking at *B*3 it is hard at first to see how the boat in the upper diagram can be heading in the direction as indicated directly below it, but nevertheless it is a fact that she is. It would almost appear that she is heading as indicated by *B*1, but by observing more closely, and noticing the position of the range lights on the upper left hand boat, it becomes apparent that it coincides exactly with the diagram of *B*3 directly below it. If such a situation is complex in a diagram, what must the actual situation appear like when you are aboard your own boat?

When only the green light is visible the situation is not unlike that just mentioned. Here again, if it were not for the range light, you could not tell from your position on *A* whether the upper boat was heading in a position indicated by *B*1, *B*2, or *B*3.

Side Lights Show Across the Bow

Too much dependence should not be placed on the supposition that the colored side lights are not showing across the bow. There are several reasons which, if not taken care of, will cause the side lights to show across the bow. The position of the lamp as a whole must necessarily be several inches at least from the inboard screen, and the width of the flame and the reflection from the after side of the lightbox all tend to make the lights show across the bow to a greater or less degree.

Special Lights

A word should be said in regard to a few of what might be called special lights. For example, a vessel not under control carries two red lights in a vertical line, one over the other, arranged to show all around the horizon. If the boat is making headway through the water she keeps up her side lights; otherwise, she does not carry them. Anchor lights are, of course, familiar to everyone. Boats of less than 150 feet in length should carry one anchor light, and boats of a greater length two such lights. A vessel aground in or near a fairway carries in addition to the regular anchor light or lights, two red lights in a vertical line, one over the other, showing all around the horizon. However, this rule applies to International Rules only, and in inland waters a vessel aground shows only the prescribed anchor lights. There are special lights for drawbridges and dredges, but in most cases these are regulated somewhat by local authorities.

Lights for Wrecks, Etc.

A vessel towing a submerged object displays her regular side lights, but instead of the regular white towing lights she displays four lights vertically arranged, the upper and lower lights being white, and the two middle ones red.

Fig. 22. Lights carried by boats at anchor. When boats are moored in a club anchorage which is not in a channel or does not interfere with traffic, no anchor lights need be carried

Steamers and other types of vessels made fast alongside a wreck or moored over a wreck which is on the bottom, partly submerged, or drifting, display a white light from the bow and stern of each outside vessel or lighter, and in addition display two red lights vertically arranged where they may best be seen from all directions. Dredges which are held in a stationary position also display a white light at each corner, and two red lights carried in a vertical line placed where they may best be seen.

CHAPTER IV

Buoys of the Various Types

(*By U. S. Lighthouse Dept.*)

BUOYS are, as a rule, employed to mark shoals or other obstructions to indicate the approaches to and limits of channels or the fairway passage through a channel, and in some cases to define anchorage grounds. There were some buoys in service at the time of the transfer of the lighthouses to the Federal Government in 1789. Buoys originally were either solid wooden spars or built up in various shapes of wooden staves, like barrels. Wooden spars are still extensively used, particularly in inside waters; but built-up buoys are now constructed of iron or steel plates.

Colors and Numbers

In order to give the proper distinctiveness, buoys are given certain characteristic colors and numbers; and, following the uniform practice of maritime nations generally, Congress by the act of September 28, 1850, prescribed that all buoys along the coast or in bays, harbors, sounds, or channels shall be colored and numbered so that passing up the coast or sound or entering the bay, harbor, or channel, red buoys with even numbers shall be passed on the starboard or right hand; black buoys with odd numbers on the port or left hand; buoys with red and black horizontal stripes without numbers shall be passed on either hand, and indicate rocks, shoals or other obstructions, with channels on either side of them; and buoys in channel ways shall be colored with black and white perpendicular stripes, without numbers, and may be passed close to, indicating mid-channels. Buoys to mark abrupt turning points in channels or obstructions requiring unusual prominence, are fitted with perches or staves surmounted by balls, cages, or other distinctive marks.

Buoys marking lightvessel stations are placed in close proximity to the lightvessel, are colored in a similar manner, and bear the letters LV, with the initials of the stations they mark. Buoys defining anchorage grounds are painted white, except those used for such purposes at a quarantine station, in which case they are painted yellow.

To assist further in distinguishing buoys, the ordinary types are made in two principal shapes in the portion showing above the waterline—nun buoys, conical in pattern with pointed tops, and can buoys, cylinder shaped with flat tops. When placed on the sides of channels, nun buoys, properly colored and numbered, are placed on the starboard or right-hand side going in from sea, and can buoys on the port or left-hand side. The numbers and letters placed on all buoys are formed by standard stencils, to insure uniformity, and the largest size practicable is used so that these may show as prominently as possible. White characters are painted on black buoys and black characters on red buoys.

Anchoring Buoys

Buoys are anchored in their positions by various types of moorings, depending on the character of the bottom and the size and importance of the buoy. They are placed in position and are cared for by the lighthouse tenders, which are provided with specially designed derricks and lifting gear for handling them. It is customary to relieve all buoys at least once a year for overhauling, repairing, cleaning, and painting, and oftener when circumstances render it necessary. Although among the most useful of aids to navigation, buoys are liable to be carried away, dragged, capsized, or sunk, as a result of ice or storm action, collision, and other accidents, and therefore may not be regarded as absolutely reliable at all times. Great effort is made, however, by the Lighthouse Service to maintain them on station in an efficient condition, which frequently requires strenuous and hazardous exertions from the crews of vessels charged with this duty. It is necessary to keep on hand at all times an ample supply of spare or relief buoys, with the necessary appendages, to provide for emergencies and the systematic relief of buoys on station.

Classes

Buoys may be divided into two general classes, lighted and unlighted, of which the latter are in the great majority. Unlighted buoys comprise spars, both wooden and iron, can, nun, bell, and whistling buoys, with a few other types for special purposes. Lighted buoys are provided with some form of gas apparatus and a lantern; frequently a bell or whistle is also attached, in which case they are known as combination buoys. A brief description of each kind follows.

Cans and Nuns

Cans and nuns, as already noted, are built of iron or steel plates, the former showing a cylindrical and the latter a conical top, and are the most extensively used of metal buoys. The interior of the buoy is divided by bulkheads or diaphragms into two or more compartments, to prevent sinking when damaged. Each kind is built in three classes or sizes, and in addition there are two general types in use—the ordinary type and the tall type, or channel buoys. The latter are a modern development of a larger and more prominent buoy for use in deeper water. These buoys weigh from 8,300 to 700 pounds each, according to size, and are generally moored by means of a stone or concrete block, or a especially designed hemispherical cast iron sinker, shackled on a length of chain about two

Fig. 24. First-class nun buoy with knife edges

Fig. 23. An acetylene gas buoy

or three times the depth of water in which the buoy is placed. The ordinary type buoys require a cast iron ballast ball attached directly below the buoy, the mooring chain being shackled in turn to the lower end of the ballast ball; this is necessary to assist the buoy in maintaining an upright position, regardless of tidal or other currents. The ballast ball is not needed with a tall type buoy, which has more stability, due to its greater draft and to a fixed counterweight of cast iron bolted on its lower end. To prevent kinking or twisting of the chain, a swivel is occasionally placed in the mooring chain for all types.

Spar Buoys

Wooden spar buoys are usually cedar, juniper, or spruce logs, trimmed, shaped, and provided with an iron strap and band at the lower end for attaching the mooring, which is as a rule a heavy stone, or concrete block, or an iron sinker, sometimes shackled directly to the buoy, or to a short piece of chain, as required by the depth. (See Fig. 25). Such buoys are among the most economical and generally used of all aids, and are particularly employed in rivers and harbors. They are, however, easily damaged by ice or collision, and in some waters suffer greatly from the attacks of the teredo and other marine borers, although this danger may be reduced by special paints or other protective treatment when not unduly expensive. Four sizes or classes are in use, varying in length from 50 to 20 feet over all, to conform properly to the depth of water at the position of the buoy. The weights of such buoys vary from 1,500 to 350 pounds each.

Iron spar buoys are built up of iron or steel plates in the form of wooden spars, and are particularly valuable where severe ice conditions exist, or where the teredo is unusually active. They are naturally more expensive and heavier to handle, thus restricting their use to special localities. They are made in three classes, in lengths of from 50 to 30 feet over all, weighing from 4,000 to 2,000 pounds, respectively.

Bell Buoys

Bell buoys have a hemispherical-shaped hull, built of steel plates, with flat deck, and carry a steel superstructure which supports a bronze bell and usually four iron clappers. The motion of the buoy in the sea causes these clappers to strike the bell, so that the action is entirely automatic. Although the buoy is quite sensitive, and responds to even a very slight motion of the waves, the sound may be faint or absent in unusual calms. This type of buoy is especially efficient in harbors or inside waters for marking points where a sound signal is desired. Bell buoys weigh about 6,900 pounds each, complete, and are moored by means of a bridle or chain attached to lugs on the opposite sides of the hull near the waterline, the same mooring

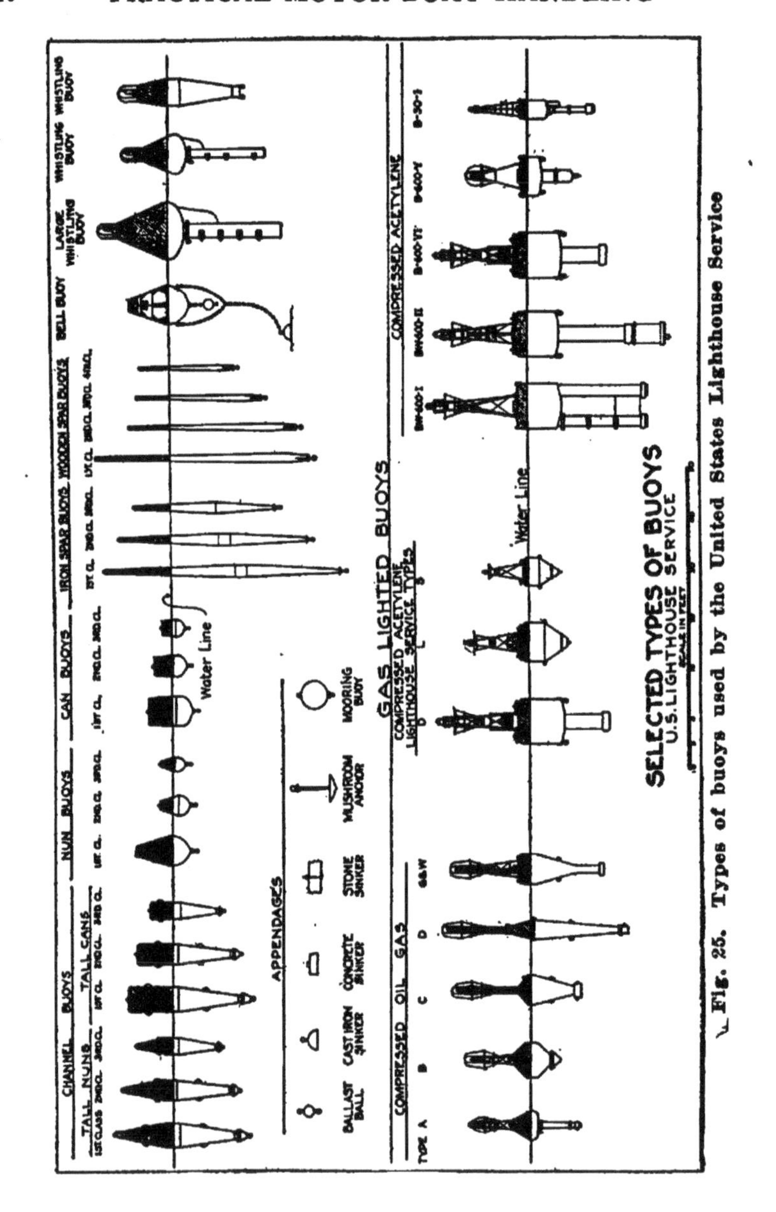

Fig. 25. Types of buoys used by the United States Lighthouse Service

being shackled to the middle and lowermost part of the bridle and extending in the customary scope of chain with a swivel to a heavy cast-iron sinker on the bottom. A large-sized ballast ball is shackled to a mooring eye at the bottom of the buoy, and the whole effect of this arrangement is to assist in the pendular motion necessary for ringing the bell.

Whistling Buoys

Whistling buoys are built of steel plates, and consist of a pear-shaped body with the smaller end uppermost, with a long open tube on the lower end. This tube extends throughout the length of the buoy, and is closed at the upper end by a head-plate on which is mounted a check valve and a whistle on the superstructure of the buoy. The sound is produced by the air in the upper portion of the tube being compressed by the falling of the buoys in the waves, its means of escape being through the whistle. A fresh supply of air is drawn through the check valve as the buoy rises again. Like the bell buoy, the sound is automatic, depending solely on the motion of the waves, and therefore the whistle may be silent when the sea is very smooth. The whistling buoy is most efficient in rough outside waters, where a ground swell exists, and is employed for important points where a sound signal is considered desirable. It is generally moored with a single chain of the proper scope and a heavy iron sinker. The weight of the buoy is about 6,500 pounds. For great depths, where the necessary quantity of chain impedes the flotation of the ordinary size of this buoy, a special and larger size is in use. This is similar to the regular size in design and operation but weighs about 11,000 pounds.

Lighted Buoys

Lighted buoys are a modern invention, having come into general use within the last twenty years, and are considered by mariners generally as among the most valuable of recent developments in coast lighting (See Figs. 23, 26 and 27). The first buoy of this kind was a gas buoy established experimentally by its manufacturers in 1881 near Scotland Lightship, entrance to New York Bay; it was officially taken over by the Lighthouse Service in April, 1884. Electric buoys, operated by a cable from shore, were established in Gedney Channel, New York Bay, in November, 1888, and were discontinued in 1903, after many mishaps, due chiefly to breaking of the cable. The operating expense was high, and in the final year of service these buoys were extinguished through accident on 120 nights.

Types of Gas Buoys

All of the lighted buoys now in service use compressed gas—either oil gas or acteylene. Various types of self-generating acetylene buoys have been in use, operating on the carbide-to-water and water-to-carbide principles, but have been abandoned on account of uncertainty of length of run, difficulty of cleaning, and danger of explosion.

In the types now in use the gas, at a pressure of about twelve atmospheres, is contained either directly in the body of the buoy or in tanks fitted into compartments of the body, and is piped to the lantern at the top of the superstructure. If the light is flashing, as is commonly the case, a small pilot light burns continuously and ignites the main burner as gas is admitted from the flashing chamber, which is a regulating compartment in the base of the lantern provided with a flexible diaphragm and valves for cutting off and opening the flow of gas at intervals, the operation being due to the pressure of the gas in the reservoirs. The length of the light and dark periods may be adjusted to produce the desired characteristic, such as five seconds light, five seconds dark, etc. Some types burn the gas as an ordinary flat flame, while others make use of an incandescent mantle, which is, however, not wholly satisfactory in rough water on account of the liability of breakage.

Reliability of Gas Buoys

Gas buoys are made in a number of different sizes, weighing from 2,800 to 34,500 pounds each, depending on the importance of the location, and burn continuously by night and day for intervals of a month to a year without recharging. The apparatus is patented by the various makers and has been brought by them to a considerable degree of perfection, so that considering the rough usage to which such buoys are subjected by the elements, gas buoys are generally satisfactory within the limits of reliability to be expected from such aids. They should not, however, be relied upon implicitly, as they may become extinguished or dragged from their proper positions, or the apparatus may be thrown out of water, some time elapsing before the buoy can be reached to repair or relight it. Gas buoys furnish valuable marks for approaching entrances, defining channels, and marking dangers, and at times may obviate the necessity for lightvessels or lighthouses on submerged sites, either of which would be many times more expensive. There is a constant demand among mariners for more gas buoys and for buoys with more brilliant lights.

CHAPTER V

Government Navigation Lights

LET us now turn our attention to what might be called Government Navigation lights, which are one branch of the aids to navigation. The United States Lighthouse Service, under whose jurisdiction the Government lights fall, is charged with the establishment and maintenance of all aids to navigation, and with all equipment and work incidental thereto on the coasts of the United States. The term "Aids to Navigation" comprises all land and sea marks established for the purpose of aiding the navigation of vessels, and includes light stations, lightvessels, fog signals, buoys of all kinds, minor lights, and day beacons.

Lighthouse Districts

The service outside of Washington is divided into nineteen lighthouse districts; each of which is under the charge of a lighthouse inspector. In each district there is a central office at a location selected either because of its maritime importance or its geographical location. Each district is provided with one or more lighthouse tenders for the purpose of distributing supplies and materials, and for the placing and care of the buoys, etc.

The jurisdiction of the lighthouse service extends over the Atlantic, Gulf, and Pacific Coasts, the Great Lakes, the principal interior rivers, Alaska, Porto Rico and Hawaii, and all other territory under the jurisdiction of the United States, with the exception of the Philippine Islands and Panama. Only one light outside of this territory is maintained wholly or in part by our Government. This light is at Cape Spartel, Morocco, and is maintained in accordance with an agreement between Morocco, the United States, Austria, Belgium, Spain, France, Great Britain, Italy, The Netherlands, Portugal, and Sweden.

The United States coast line, including the Philippines, Panama, the Great Lakes and the rivers under the jurisdiction of the Lighthouse Service has a length of 48,881 miles. Omitting the coast line of the Philippines and Panama, we have a net mileage under the jurisdiction of the service of 37,381.

Number of Aids to Navigation

Fig. 29 shows a summary of the aids to navigation under each principal class in commission on June 30, 1915. It will be noted that there was a total of 14,544 aids on that date.

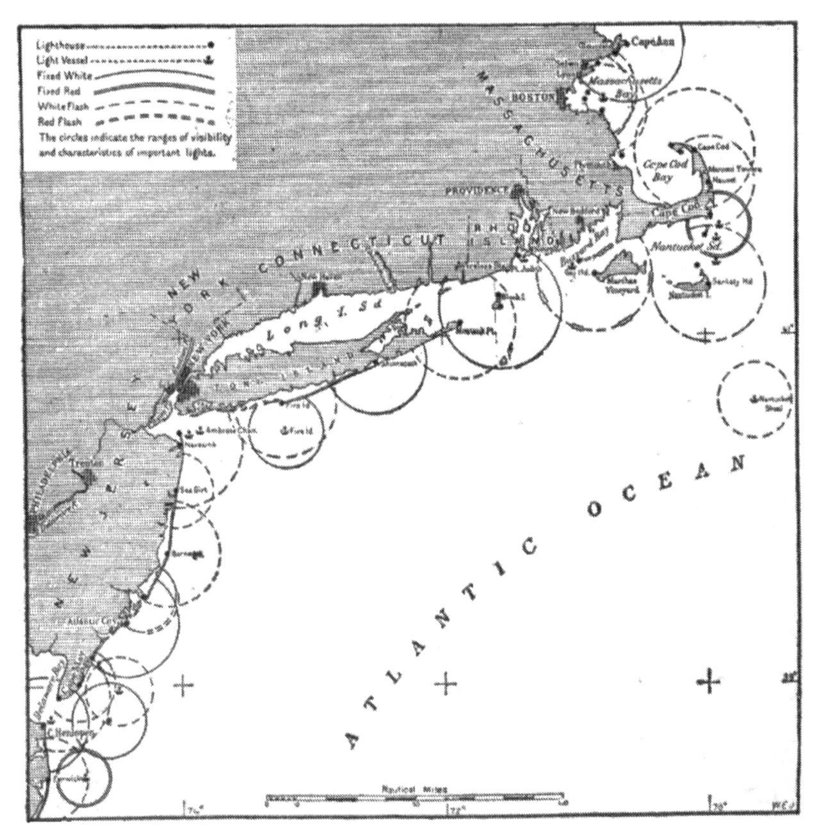

Fig. 28. Method of lighting a portion of the Atlantic Coast by means of major lights—their range of visibility

The term "minor lights" includes post lights and small lights which are generally not attended by resident keepers. These lights are usually cared for by persons living in the vicinity who are not obliged to devote their entire time to the work. Lightvessels are used to mark offshore dangers, or the approaches to harbors or channels where lighthouses would not be feasible or economical. Gas buoys are used to mark harbor channels or shoals. Float lights are usually small lights borne on a float or raft; they are employed for less important places where more convenient or economical than lighted buoys. Fog signals include the various types of aerial sound-producing apparatus for use in foggy or thick weather. They embrace various types of whistles, sirens or horns, actuated by steam or compressed air, and bells operated by machinery or hand.

Lighted Aids	
Lights (other than minor lights)	1,662
Minor lights	2,837
Lightvessel stations	53
Gas buoys	479
Float lights	124
Total	5,155
Unlighted Aids	
Fog signals	527
Submarine signals	50
Whistling buoys, unlighted	86
Bell buoys, unlighted	237
Other buoys	6,488
Day beacons	2,001
Total	9,389
Grand Total	14,544

Fig. 29. Total number of aids to navigation

Early History

The history of lighthouses in the United States dates back to 1715, when the first lighthouse on this continent was built at the entrance of Boston Harbor by the Province of Massachusetts. The light was supported by light dues on all incoming and outgoing vessels except coasters. Several other lighthouses were built by the Colonies before 1789, when Congress authorized that the lighthouses and other aids to navigation be maintained at the expense of the United States. The lighthouse service of the United States is now supported entirely by appropriations out of the general revenues of the Government, and the United States lighthouses have been free to vessels of all nations from 1789 to the present time. There is no system of light dues, as is the case in a number of foreign maritime countries.

Lighthouse keepers receive a yearly salary of from $600 to $1,000, depending upon the importance of the light, etc. Attendants of post lights receive on the average $10 per month per light. Each large lighthouse tender costs the service about $40,500 annually, a lightvessel $15,300, and an important light station, with fog signal, $4,200. The average maintenance on a gas buoy is $100 to $300.

Lighting Apparatus

The earliest type of lighting apparatus consisted of an open coal or wood fire, with other inflammable materials such as pitch burned on top of a tower. When Boston Light was established in 1716 the common oil burner of that period was used, enclosed in a lantern consisting of a cylinder of heavy wooden frames, holding small thick panes of glass. The illuminant was fish or whale oil. Sperm oil was in general use about 1812, and was burned in a lamp with a rough reflector, and a so-called lens or magnifier. Improvements were gradually made in this apparatus, and by the year 1840 the useless bull's-eye magnifiers had been entirely removed, and the reflectors were made on correct optical principles. To provide illumination all around the horizon, sets of from eight to twenty lamps were used, placed side by side around the circumference of a circle. The first lens in the United States was installed at Navesink Light, N. J., in 1841, and is still preserved by the service.

During the transition period of lighthouse apparatus from reflectors to lenses, sperm oil remained as the leading illuminator, until its price made its use prohibitive. Colza oil was used in small quantities about 1862, but during the period from 1864 to 1867 lard oil was adopted as a standard illuminant, and was generally employed until 1878, when kerosene came into use. Its use gradually increased, and about 1884 kerosene had become the principal illuminant, and so remains at the present time. The lamps used were also improved, passing through various styles to a special form of concentric wick, using five wicks for the larger sizes. The incandescent oil-vapor lamp, which is now generally employed for important lights, burns vaporized kerosene under an incandescent mantle, giving a much more powerful light, with little or no increase in consumption.

Various other illuminants are now in use. Oil gas is extensively used, particularly for lighted buoys, and acetylene gas is employed for light buoys and unattended light beacons. Electric ar and incandescent lights are used in special instances. Electric lights with distant control are employed in a number of cases where a reliable source of current can be obtained.

Classification of Lights

Lights have heretofore been classed according to their order; that is, first order, second order, third order, etc., down to the sixth order, inclusive. The order of the lens depends upon the inside radius or focal distance of the lens, that is, the distance from the center of the light to the inner surface of the lens. In a first order light this distance is 36.2 inches, in the second order 27.6 inches, and in the sixth order light the focal distance is 5.9 inches. The power of light does not vary directly with the order. The designation of lights by orders has, therefore, been discontinued in the light lists, and instead the candlepower is given. From the stated candlepowers the mariner may judge relative brilliancy and power of the various lights. Candlepowers are stated approximately in English candles, but the intensity of the lights as seen from a boat may be greatly lessened, or a light made invisible by unfavorable conditions, due to haze, fog, rain, or smoke.

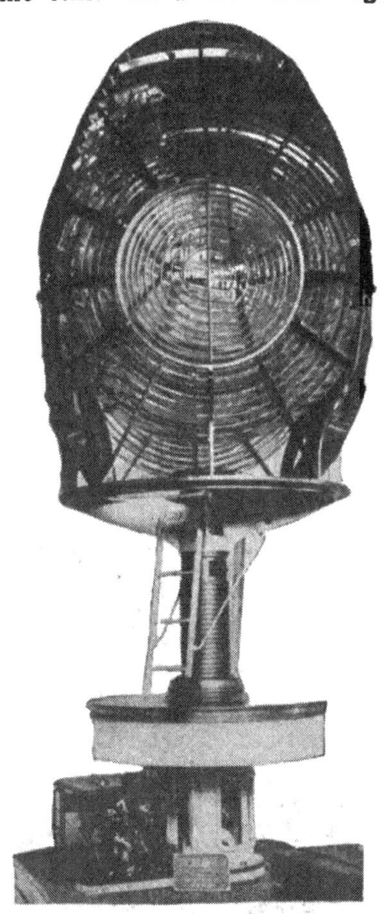

Fig. 30. First order lens used by the Lighthouse Department

Range of Visibility

Under normal atmospheric conditions, the visibility of the light depends upon its height and intensity, the distance due to the former being known as a geographic range, and to the latter as a luminous range. As a rule, for the principal lights, the luminous range is greater than the geographic range—that is, the distance from which the principal lights are visible is limited by the horizon only, and in some atmospheric conditions the glare of the light, and occasionally the light itself, may be visible beyond the computed geographic range. The distances of visibility are given in nautical miles.

Characteristics of Lights

In order to avoid the likelihood of confusion between lights, endeavor is made to give them distinctive characteristics. Since much of the coast was lighted before the introduction of modern lighthouse apparatus, the original lights were as a rule fixed, but at the more important of these stations apparatus has now been installed to make the lights flashing or occulting. This effect is produced in the case of flashing lights by revolving all or part of the lens, and in the case of occulting lights by some form of traveling screen or shutter, which obscures the light at intervals. In either case the regulation is by clockwork.

Fig. 31. The old and new lighthouses at Cape Charles

The usual phases are as follows:

Fixed: Showing a continuous steady light.
Flashing: Showing a single flash at regular intervals.
Fixed and Flashing: Showing a fixed light varied at regular intervals by a single flash of greater brilliancy.
Group Flashing: Showing at regular intervals groups of flashes.
Occulting: Showing a steady light, suddenly and totally eclipsed at regular intervals.
Group Occulting: Showing a steady light suddenly and totally eclipsed by a group of two or more eclipses at regular intervals.

The above refers only to lights which do not change color, commonly white, but further diversification is obtained by

Fig. 32. Showing the characteristic day mark used on the lighthouse at Cape Hatteras

the use of red screens, changing the color from white to red in various combinations. Such lights are known as alternating. In the case of gas or electric lights the supply of gas or current is cut off at intervals.

The term flashing or occulting refers to the relative duration of light and darkness, the flash being an interval shorter than the duration of an eclipse, and occultation being shorter than or equal to the duration of light. Red sectors are produced by screens of colored glass. They are often employed to mark outlying dangers near the light or the limits of channels, and are usually arranged so that the light shows white while a passing vessel is clear of such dangers, changing to red as a shoal or other obstruction is approached.

Day Marks

To assist identification in daylight, towers are frequently distinguished by characteristic painting, in addition to peculiarities of form or outline. The effect of certain colors when combined in bold patterns of spirals, bands or blocks is quite striking in a number of important lighthouses. Fig. 32 shows the characteristic markings of a lighthouse at Cape Hatteras, N. C. This tower is the tallest in this country, being 200 feet high. The light is visible from the deck of a vessel twenty nautical miles distant. Its characteristic is a flashing light for ten seconds.

Day Marks for Vessels

A vessel towing a submerged object in the daytime shows two shapes, one above the other in the form of a double frustum or cone, base to base, the upper cone being painted with alternating horizontal stripes of black and white, and the lower shape being painted bright red.

Steamers, lighters and other vessels made fast alongside a wreck, or moored over a wreck, display two shapes similar to the foregoing, except that both shapes are painted bright red.

Dredges held in a stationary position show two balls in the daytime, vertically arranged, and placed in a position where they can best be seen.

Self-propelled suction dredges under way, with their suction on the bottom, display the same signals as used to designate a steamer not under control, that is, two black balls placed where they may best be seen from all directions.

Vessels which are moored or anchored and engaged in laying pipe or operating on submarine construction, display in the daytime two balls in a vertical line, the upper ball being painted with alternating black and white vertical stripes, and the lower ball being bright red.

CHAPTER VI

Equipment Required by Law

Class I—Boats Under 26 Feet, L. O. A.

Lights—Combination red and green lantern (or bow and colored side lights) and stern light.

Sound Apparatus—Whistle capable of producing blast prolonged for at least 2 seconds.

Class II—Boats 26-40 Feet, L. O. A.

Lights—White forward light (lens at least 19 sq. in.); white stern light; green starboard light; red port (lenses at least 16 sq. in.); screens at least 18 in. long; lenses, fresnel or fluted glass.

Sound Apparatus—Same as Class I plus fog-horn and bell.

Class III—Boats 40-65 Feet, L. O. A.

Lights—White forward light with with lens at least 31 sq. in.; white stern light; green starboard light; red port light (lenses at least 25 sq. in.); screens at least 24 in. long; lenses, fresnel or fluted glass.

Sound Apparatus—Same as Class II, except bell must be at least 8 in. across mouth.

All Classes

One life preserver for each person on board. [Life preservers, life belts, buoyant cushions, ring buoys, or similar devices in sufficient number for every person on board, and placed so as to be readily accessible. Life presesvers or buoyant cushions must be capable of keeping afloat for 24 hours a weight exerting a direct downward pull of 20 pounds, on boats not carrying passengers for hire. No pneumatic life-saving appliances, or appliances filled with granulated cork will be permitted. Planks, gratings, etc., or small boats in tow cannot be substituted for required life-saving appliances. Floats of seasoned wood, not exceeding white pine in weight and measuring at least 4 feet by 14 inches by 2 inches, may be used.]

A fire extinguisher capable of extinguishing gasoline fires.

At anchor, a white light only, less than 20 feet above hull, visible around horizon for at least one mile.

Two copies of the Pilot Rules must be carried on board.

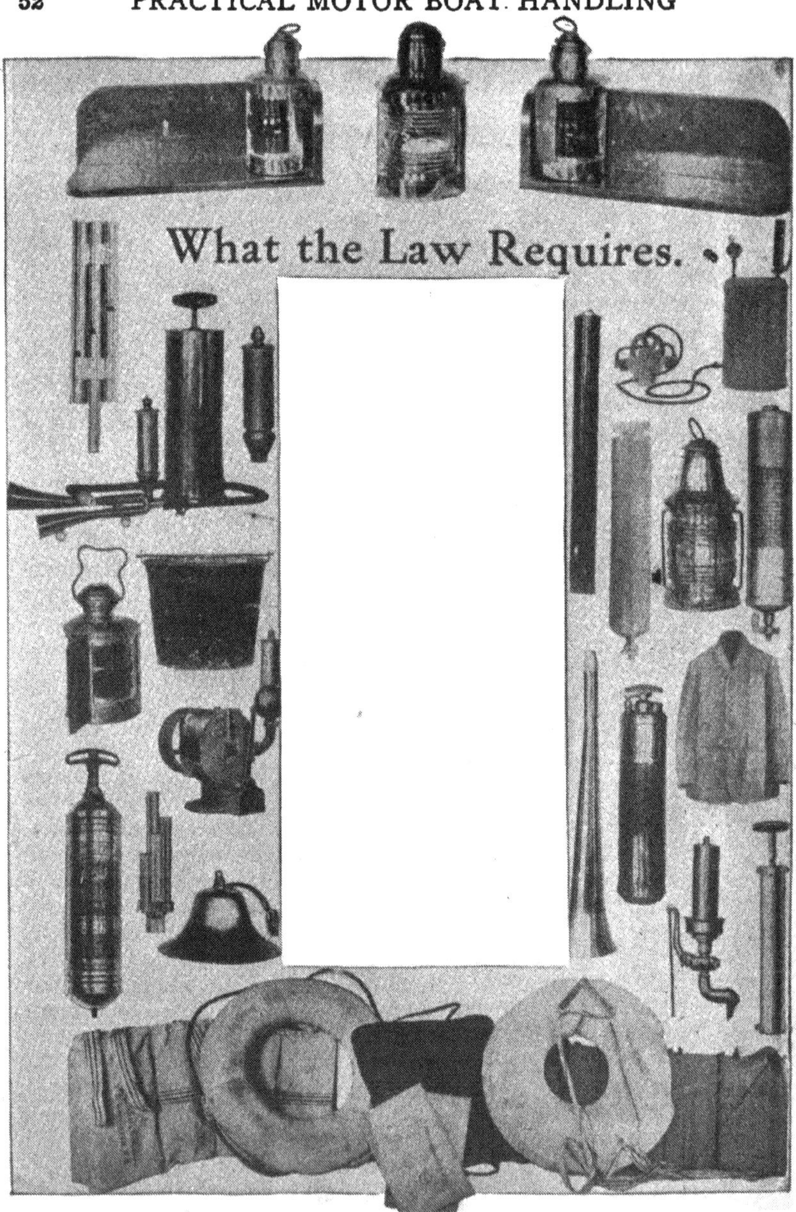

Fig. 33. Some of the equipment required by law which has been approved by the Government

CHAPTER VII

The Compass

THE discovery and early history of the mariner's compass is extremely doubtful, the Chinese, Arabs, Greeks, Finns and Italians all having been declared its originators. There is now little doubt that the claim formerly advanced in favor of the Chinese is ill founded. There is no genuine record of a Chinese marine compass before A. D. 1297. No sea-going ships were built in China before 139 B. C.

What the Compass Is

The compass is nothing more than a magnet suspended so as to be allowed to swing freely in a horizontal plane. In theory, an ordinary knitting needle magnetized by drawing a toy magnet along its length a few times, and suspended from the center by means of a thread so that it can swing in a horizontal plane, is as much of a compass as the ones we use on our boats to-day. If such a needle is magnetized and suspended it will immediately assume a north and south position.

But the compass consists of a number of magnetized needles bound together, and suspended or pivoted from beneath. On this bundle of magnetized needles we have a card mounted to give us a better sense of direction, and allow us to determine directions other than north and south, which would be the only two indicated by the magnetized needles if we had no card mounted thereon.

There has been little or no change in the mariner's compass for centuries. In theory and construction it is practically the same as it was more than one hundred years ago. The only changes which have been made are refinement in its construction, and the markings on the compass card.

The Dry Compass

The older compasses were known as dry compasses; that is, simply magnetic needles and a card pivoted at the center. Naturally such an arrangement was very sensitive and responded to the motion of the ship very freely. With the coming of the steam engine, and later, the internal combustion motor, it was found that the vibrations set up by the machinery were such as to keep the compass card in constant motion, which naturally made it unreliable as a navigating instrument. The development of the liquid compass followed, and this type overcomes to a large extent the difficulty and trouble experienced with the dry compass.

The Liquid Compass

The liquid, or wet compass, is practically no different from the dry compass, with the exception that a liquid generally consisting of a mixture of 55 per cent. water, and 45 per cent. alcohol is introduced into the bowl of the compass, and then the latter is sealed up. The liquid not only prevents the compass needle and card from responding to small vibrations due to power plants and the sea, but also tends to buoy up or float the needle and card, and thus make it rest more lightly on its pivot. This allows the card to turn more freely as the ship is turned, or rather to hold its position more steadily as the ship's bow is turned away from the compass. The smaller and less expensive compasses use kerosene as the filling liquid, and some of the newer makes use oil instead of alcohol and water. On account of the nature of the various kinds of fluids used, the compass as we know it is practically non-freezable in ordinary latitudes.

The Lubberline

Compasses are fitted with a gimbal ring to keep the bowl and card level under every circumstance of a ship's motion in a seaway, the ring being connected with a binnacle or compass box by means of journals or knife edges. On the inside of every compass bowl is drawn a vertical black line called the lubberline, and it is imperative that the compass be placed in the binnacle or on the boat so that a line joining the pivot and the lubberline shall be parallel to the keel of the boat. Thus, the lubberline always indicates the compass direction on which the boat is heading.

The Old Card

Generally speaking, there are two methods in use for marking or dividing the compass card, which we may designate for want of better names as the old card, and the new card. On the old card shown in Fig. 34 it will be noticed that the card is divided into 32 major divisions known as points, and that those major divisions are further subdivided into four parts. It will also be observed that the card is divided on its periphery into degrees. It is in the method of putting the degrees on the compass card that the new card differs from the old one.

Degrees

On the old card North and South are both marked zero, and East and West are each marked 90, the divisions running from North and from South towards East and West from zero degrees to 90 degrees. In other words, we have 45 marked opposite Northeast, as well as Southeast, Southwest and Northwest. To steer a course by this method of dividing the card it is necessary to add either the designation North or South to the degrees; that is, if we wished to steer Northeast we should call our course North 45 degrees East, and if we wished to steer Southeast this course would be called South 45 degrees East. Similarly, Southwest is

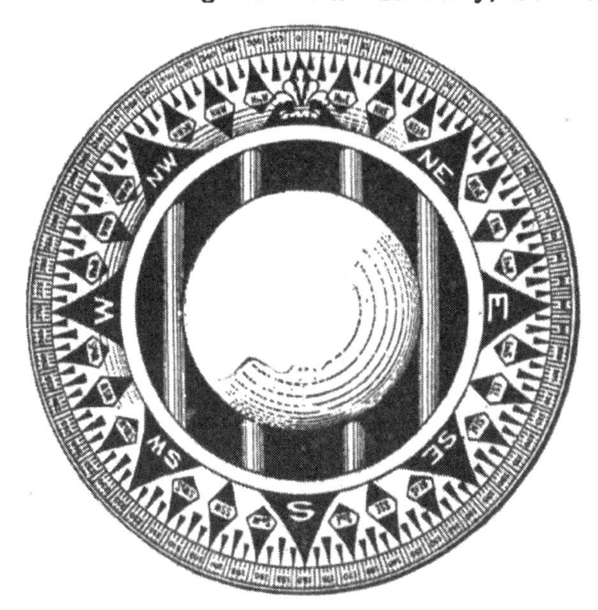

Fig. 34. The compass card

South 45 degrees West, and Northwest North 45 degrees West. One advantage of this method of designating the compass is the ease with which reverse courses may be remembered. For example, if a certain course were North 60 degrees East, then when returning over the same course the compass would indicate South 60 degrees West.

Points

Division of the compass card according to points, is a most interesting one, and as it is the method which is generally used by motor boatmen it is the one which will require our greatest attention. As mentioned above, the card is divided into 32 major divisions known as points, each one of these points having a particular name. The four principal, or cardinal points are known as North, South, East and West. The inter-cardinal points are the ones midway between the cardinals, and these are given a name which is a combination of the points which they bisect; that is, the point midway between North and East is known as Northeast, etc. This gives us eight divisions. We now subdivide these eight divisions in half, and once again we give these eight new points names which are combinations of the two points which they are midway between. For example, the point midway between North and Northeast is North Northeast. That point midway between South and Southwest is South Southwest. To get the additional 16 points it is simply necessary to divide points which we have already determined in a similar way as before. Here again the new points will have names corresponding to the points to which they are adjacent. The word "by" will be used in all of these 16 new points. For instance, the point between North and North Northeast is known as North by East, because it is adjacent to North, and in an easterly direction from North. The point between Southeast and South Southeast is known as Southeast by South, because it is adjacent to the inter-cardinal point Southeast, and in a southerly direction from it.

Quarter Points

For the purpose of steering more accurate courses than would be possible by following only 32 points, we must subdivide the points into halves and quarters. The naming of these quarter points is most interesting, and must be thoroughly mastered by the motor boatman at the beginning. Naturally it will be seen that every quarter point might have two names; that is, it might refer to the point either to the right or to the left of it. For example, the quarter point just to the right of North could logically be called North ¼ East, or it might be called North by East ¾ North. Either of these designations would probably convey to the man at the wheel the course which it was desired that he should follow. However, and perhaps unfortunately, there is a certain method of calling these quarter points, and again we are confronted with two methods instead of one.

The older method appears to many to be the most logical one, although the Navy Department has seen fit to adopt one of its own. It makes little difference which of these methods is adopted by the motor boatman. Both are correct.

Fig. 35 shows the two methods of naming the quarter points, and in it Fig. I shows the older method. This system is to name the quarter points from each cardinal or inter-

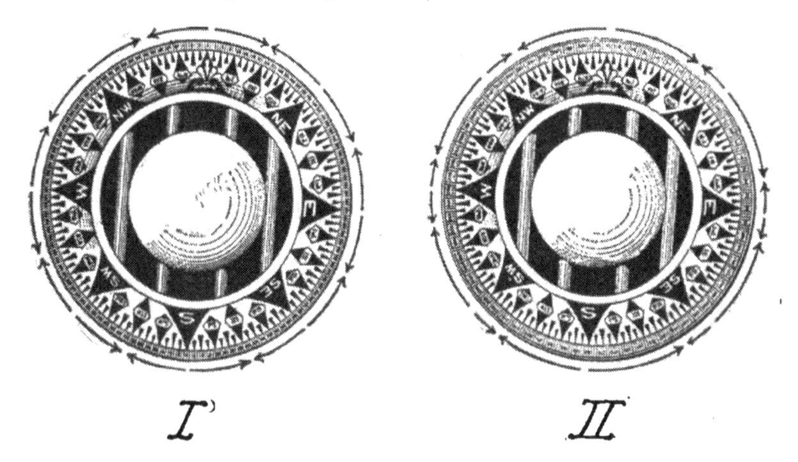

Fig. 35. The two methods used in naming the quarter points of the compass

cardinal point toward a 22½-degree point; that is, toward North Northeast, East Northeast, East Southeast, South Southeast, South Southwest, West Southwest, West Northwest, and North Northwest. The United States Navy method is to name the quarter points from North and from South towards East and West, excepting that the division adjacent to a cardinal or inter-cardinal point is always connected with that point. This method is shown in II

CHAPTER VIII

Compass Errors

EVERYONE knows that the compass points North, or at least, should point North. Unfortunately, we have two Norths. One of these is the upper extremity of the earth's axis, and is known as the geographic or true North. The compass does not point to this North, but always points towards what is known as the magnetic North. Magnetic North is located at some distance from the true North, roughly indicated by Fig. 36.

Variation

If you were on your boat at the position marked *A*, and your boat was heading as indicated by the dotted line, she would be heading true North, but the compass would be pointing in a decidedly different direction, indicated by the dotted line from *A* with a point marked *MN*. In other words, your boat would be headed true North, but the magnetic heading would be quite different. This angular difference between the true North and the magnetic North is known as the variation of the compass, shown in Fig. 36 by the angle between *TN*, *A*, and *MN*.

One will immediately see from Fig. 36 that this variation of the compass is not constant; that is, it is different with every change in geographical location. If your position is at *B*, your boat is still heading true North, and your compass towards *MN*. Observe that the angle between true North and the magnetic North at position *B* is decidedly different and smaller than when at *A*. In other words, the variation of the compass at *B* is much less than at *A*. In both cases the magnetic North has been to the West, or to the left of the true North, which makes the variation what is known as westerly. Now consider for a moment your position at *C*. In this case you will notice that your boat is heading towards the true North, and also towards the magnetic North. In other words, there is no angle between the two poles. Therefore, at position *C*, or anywhere along the dotted line leading from *C* towards the poles, the variation is zero. At *D* we again have a variation, but in this case the magnetic North is to the East, or to the right of the true North, and we, therefore, have an easterly variation. At *E*, the boat is heading towards the true North, but going away from the magnetic North. In such a case, while the boat is heading North, the compass is pointing South, and we have 180 degrees variation.

Change in Situation

As has just been mentioned, the variation of the compass is different for every geographical location. In the vicinity of New York City the variation is about 9 degrees westerly; around Portland, Me., it is about 15 degrees westerly. As we go West the variation becomes less and less until in the vicinity of Lake Superior we have zero variation. Farther West than this the variation becomes easterly, and increases in magnitude.

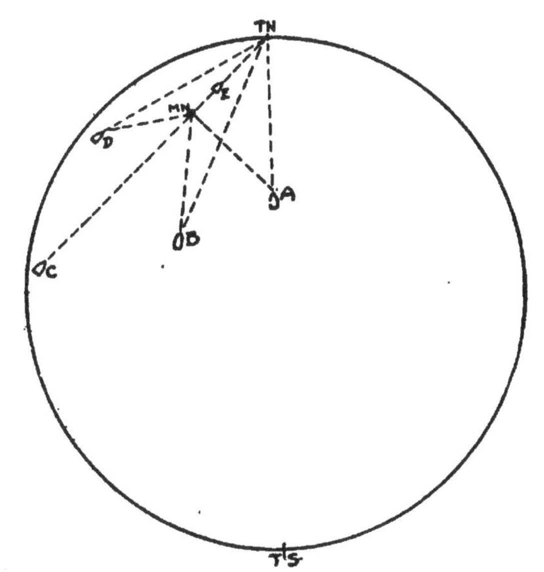

Fig. 36. Showing variation of the compass and how it differs in amount at different locations on the earth

Determining Variation

One may now ask how to determine what this variation is. This is a relatively simple matter, as on every one of our Government charts this information is given. Fig. 37 gives what is known as the compass rose on the chart, several of which are printed on every one. The note in the center of this rose, "Variation 13 degrees 40 minutes West in 1915," gives the information in regard to variation at the particular location where this rose is printed. From the statement directly below, "Annual increase 6 minutes," it will be recognized that variation is not a constant quantity, but is increasing or diminishing all the time. To calculate what the variation is to-day, that is, in 1917, we simply must add 12 minutes to the variation as noted above.

On this compass rose there are an inner and an outer set of divisions, the inner one being in points and quarter points, and the outer in degrees. The two do not correspond—that is the North magnetic division is not pointing to zero degrees but to a division about 14 degrees West of zero. The explanation of this is that the outer divisions in degrees refer to the true North, and are known as true courses, while the inner divisions refer to the magnetic North. It is almost invariably true that when courses are given in degrees they are true courses, and that when they are given in points they are magnetic courses. The magnetic ones are far the more simpler for our use, as we need to take no account of variation whatsoever in dealing with magnetic courses. Variation comes in only when we refer to true courses.

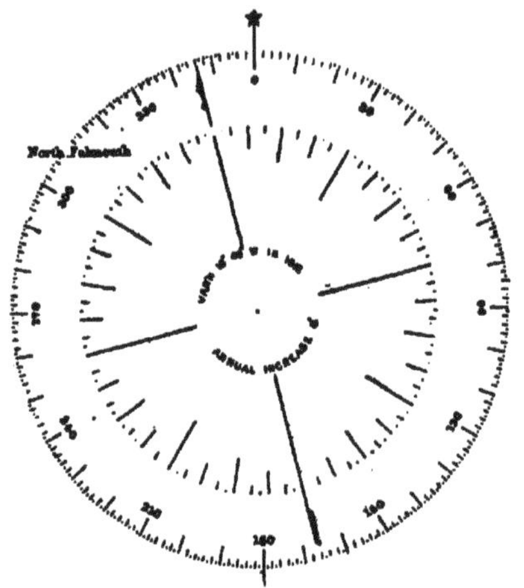

Fig. 37. The compass rose showing the amount of variation. Several of these roses are to be found on every chart

Deviation

But there is one error entering into our compasses both when we talk about magnetic, as well as when referring to true courses, and this error is caused by magnetic substances such as iron and steel on our boats. The error caused by the effect which this magnetic substance has on our compass, moving the needle one way or the other, is called deviation. It exists to a greater or less degree on every motor boat. Moreover, deviation on any boat is not constant; that is, it is different in amount for every different heading of a boat.

Fig. 38 shows why this difference in the amount of deviation occurs. Here we have three boats. In the first case, the boat is heading approximately North. The black dot is used to represent the center of magnetic attraction on the boat. When the boat is heading approximately North, as shown, the pull of this center of magnetic attraction will be exerted most strongly on the South point of the compass, and in the direction which is approximately Southeast. As the boat swings around to the easterly direction, it is apparent that the attraction is on altogether different points of the compass. Naturally, this will cause the compass to have a deviation decidedly different from that of the first case.

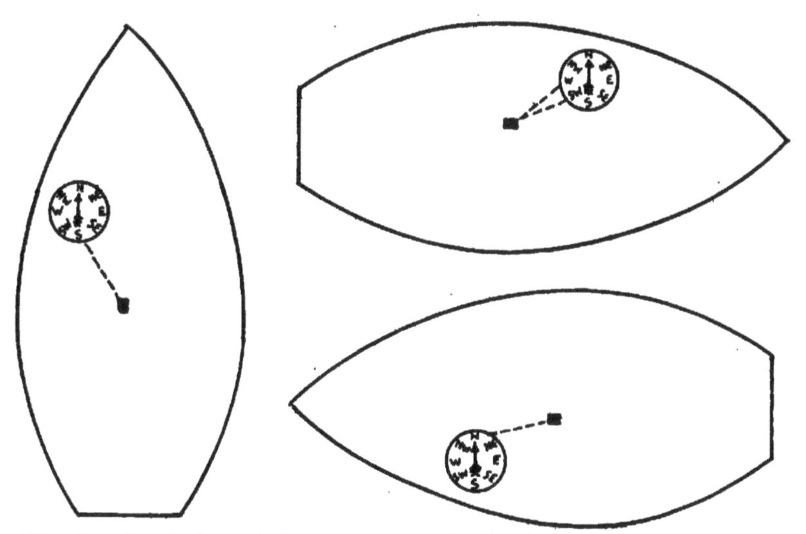

Fig. 33. Deviation of the compass and why it differs on the various headings

As the boat swings around to a westerly direction, the pull of the magnetic substances on the boat is again different, and causes an entirely different effect on the compass. In other words, the deviation is different on every different heading. It cannot be assumed that because we have one point westerly deviation when heading North we will have the same amount when heading East or South.

Determining Deviation

The question now arises not only how to determine the deviation of one's compass, but how to correct the compass so that no deviation will exist. The former is simple, the latter, very complex and difficult. It is much simpler to determine the deviation, know how much it is, and let it exist, than to attempt to correct and compensate for this error.

To determine the amount of deviation, it is simply necessary to choose a number of courses whose direction can be determined from one's chart, and then put one's boat over these courses, and note the direction shown by the compass. For example, choose two points on the chart which are directly North and South of each other; that is, two lighthouses, buoys, headlands, or other points which can be readily distinguished. Put the boat over this course, and note the course which the compass shows. Perhaps it will be North by East. Make a note of this. Now turn the boat directly about, and she will be heading in a southerly direction. Again note what the compass shows, and set it down on paper. In this way pick out as many courses as possible, and put your boat over them, noting in each case what the compass shows, which will give you a deviation card.

Easterly and Westerly Deviation

When the north pole of your compass is swung to the right, or toward the East by the magnetic substance on the boat, the deviation is said to be easterly. When the north pole is swung to the left or to the West, we have a westerly deviation. Deviation refers to the north point of the compass, and to no other point—which fact should be remembered by everyone.

Applying Deviation

The process of applying deviation to determine compass courses is one which the navigator must do for himself, and make himself a thorough master of. No course can be set or bearing plotted without the application of this problem, and a mistake in its solution may produce serious consequences. Rules as to the application of deviation are of little service. The motor boatman must practise and work them out for himself.

Compass courses and magnetic courses should not be confused. The former is that shown by the compass on your boat, and the magnetic or correct course is the one shown by the chart. To find a compass course when the deviation of your compass is westerly, the compass course which you should steer will be to the right of the magnetic or correct course. In other words, apply a westerly error to the right to find the compass course which should be steered. When the error of your compass is easterly, the compass course which should be steered to allow for this easterly error, is to the left.

To find the magnetic, or true course from your compass just the reverse of the above must be done; apply an easterly error to the right, and a westerly error to the left.

Fig. 39 shows three boats heading in exactly the same direction. In Fig. I of this diagram, there is zero deviation. In this case the magnetic course is N N W, the true course N N W, and the compass course N N W. In Fig. II, we

Fig. 40. Determining deviation by means of the sun compass

have a variation of two points westerly, but with zero deviation. In this case the true course is N N W, the magnetic course is N, and the compass course N. In Fig. III we have two points westerly variation and one point easterly deviation. Here we have a true course of N N W, a compass course of N by W, and a magnetic course of N.

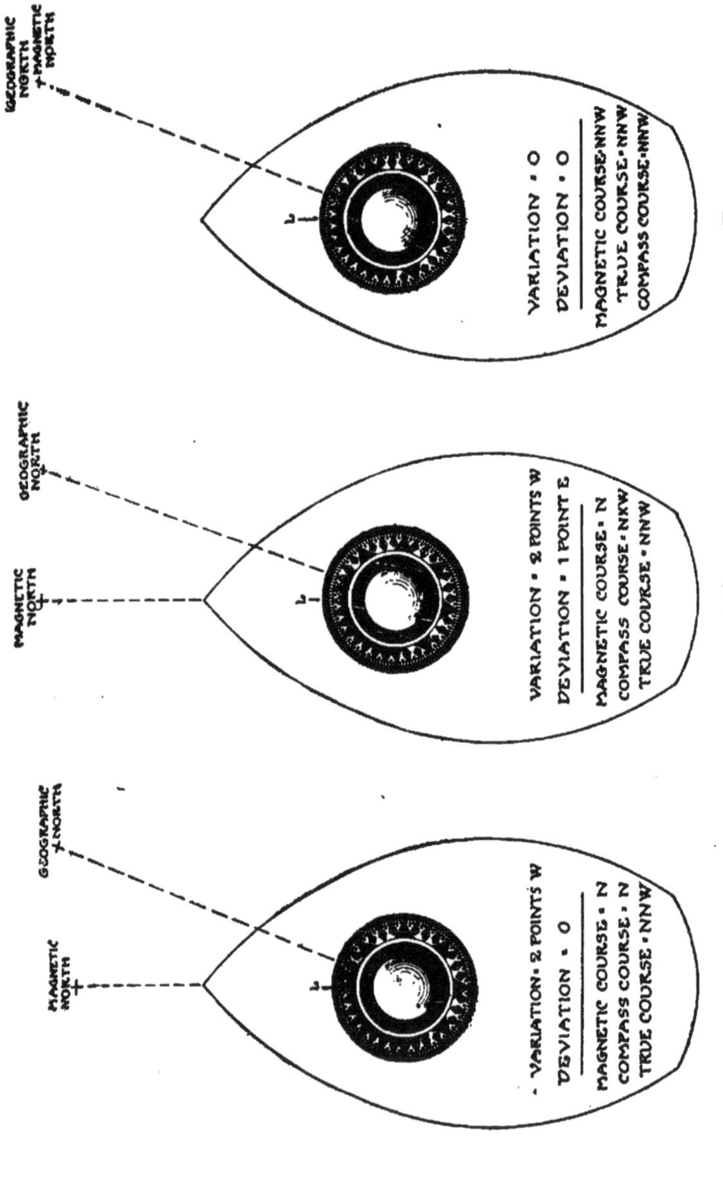

Fig. 39. Examples of true, magnetic, and compass courses with various amounts of variation and deviation of the compass

The Sun Compass

If one does not look with favor upon the scheme of determining the deviation of his compass on various headings, then probably the next best method is that which employs a sun compass or shadow pelorus, as it is sometimes called. Any amateur without previous experience can determine the devia-

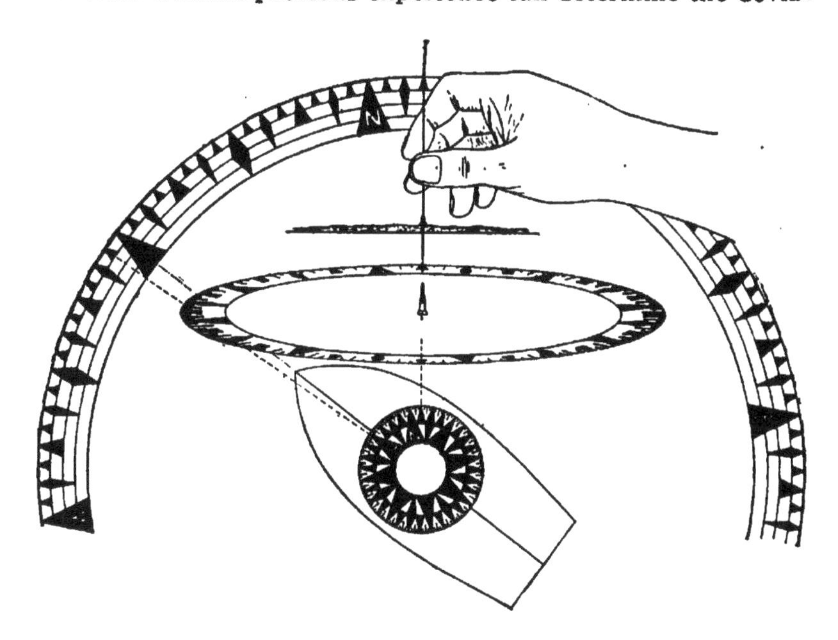

Fig. 41. Determining deviation by means of a bearing on a distant object, a brass screw and a piece of wire. The outside semicircle is included in the diagram merely as an indication that the lighthouse (shown with a strip of shore line immediately below the hand) bears N ¾ E magnetic from the boat's position. The inner (elliptical) circle represents the boat's compass card with the lighthouse bearing N by E when the boat is heading NW, thereby revealing a deviation of ¼ point westerly. The compass course is, therefore, NW ¼ N.

tion of his compass within a quarter point with the sun compass. As will be seen from Fig. 40, the sun compass is nothing more nor less than a reversed compass card, so mounted in a box that it can be turned around its center at will. At its center a hole is drilled, and a straight wire some three or four inches in length projects vertically upward. On the box at *A* a line corresponding to the lubberline of the magnetic compass is cut, and the sun compass is always so placed on the boat that the mark *A* will represent the bow of the boat either by being in line with the bow if the sun compass is in line with the keel of the boat, or else so that an imaginary line

Fig. 42. A deviation card

drawn through *A* and the center of the compass will be parallel to the keel of the boat.

After the sun compass has been placed as noted above, it is an easy matter to determine the heading of the boat. By comparing this heading with that heading as indicated by the boat's compass, the deviation can be read off directly. There is a small table which accompanies the sun compass that must be used with it in determining the deviation. This table tells one at what figure on the sun compass the line at *A* must be set for any time of the day. The sun shining on the sun compass causes a shadow to fall from the upright wire, and this shadow cuts a point indicating the heading of the vessel. For example, on May 1 at 10.20 A. M. at New York City the table will tell one to set the movable dial at 115. If the shadow from the wire then falls for example at S S E 3/8 E, this will be the true direction in which the boat is heading. Now, if the compass shows S by E at the same moment, and we have a variation of one point as shown by the chart, we know at once that our compass has an error due to deviation of 3/8 of one point.

The Deviation Card

Fig. 42 shows another form of deviation card which is very convenient. The inner compass is magnetic, and the outer one represents the compass on your craft. Put your boat over a number of courses whose magnetic direction can be determined from the chart, and note the headings as indicated by the compass. Draw a line in each case on the above card from the point on the inner (magnetic) compass, representing the chart course, to the outer point which is indicated by your compass. Going over eight courses and the reverse of them will give you the deviation of your compass on sixteen different headings. You will then have a deviation card which will show the compass course which should be steered for any magnetic course.

CHAPTER IX

The Chart

THE nautical chart is a miniature representation of a portion of the navigable waters of the globe. It generally includes an outline of adjacent lands, aids to navigation, depth of waters, character of the bottom, etc., with considerable information in tables. The chart should be carefully studied, and among other things all of its notes (See Fig. 43) should be read, as valuable information may be given in the margin which it is not practicable to place upon the chart abreast of the locality affected.

The motor boatman should be especially careful that the chart is of recent issue, or bears corrections of recent date, which facts should always be clearly shown upon its face. It is well to proceed with caution when the chart of any locality is based upon an old survey. Even if the original sur-

SOUNDINGS

The soundings are in fathoms except on the tinted surfaces, where they are in feet, and show the depth at mean low water.

SIGNS AND ABBREVIATIONS

✦ *L.S.S. life saving station.* (T) *connected with general telegraphic system.*
✦ *Stations of Mass. Humane Society*

C. can, N. nun, S. spar.
Red buoy; to be left to starboard in entering;
Black buoy; to be left to port in entering.
Black and red horizontal stripes; danger buoy
Black and white perpendicular stripes; channel buoy.

** Rock awash at any stage of the tide.*
+ Sunken rock. *Wreck.*

M. mud, S. sand, G. gravel, Sh. shells, P. pebbles, Sp. specks,
bk. black, wh. white, rd. red, yl. yellow, gy. gray, bu. blue, dk. dark, lt. light
hrd. hard, sft. soft, fne. fine, crs. coarse, rky. rocky, stk. sticky, brk. broken,

Fig. 43. A note from a Government chart showing some of the information given on charts

LIGHTS

F. signifies Fixed, Flg. Flashing, Fl. Flash, Fls. Flashes, Rev. Revolving, W. White, R. Red, v. varied by Sec. Sector.

Name	Latitude	Longitude West from Greenwich	Character	Interval bet. flashes	Color of structure	Height above sea	Visibility in naut. miles at 15 ft. elevation	Fog signals	Submarine Bells
Cross Rip (Light Vessel)			F.R.		Black	39 ft.	11½	Bell or Horn	
Hyannis (Rear Range)	41°38'11"	70° 17'20"	"		White	42 "			
Hyannis Bn. (Front ")			"		Lead	22 "			
Cape Poge	41°25'16"	70°27'06"	Flg.W.&R.	0m05s	White	54 "	12¾		
Edgartown	41°23'27"	70°30'13"	F.W.		"	48 "	12¼	Bell	
East Chop	41°28'13"	70°34'05"	Flg.R.	0m10s	Brown	79 "	14½	"	
West Chop	41°28'51"	70°36'01"	F.W. R.Sec.		White	83 "	14¾	Whistle	
Nobska Point	41°30'57"	70°39'20"	" "		White	86 "	15	Bell	
Tarpaulin Cove	41°28'08"	70°45'29"	F.W.v.W.Fl.	0m30s	White	77 "	14¼	"	
Gay Head	41°20'55"	70°50'08"	Flg.W.&R	0m10s	Red	170 "	19¼		
Cuttyhunk	41°24'52"	70°57'01"	F.W.		White	61 "	12¾		
Dumpling Rock	41°32'18"	70°55'19"	F.W. R.Sec.		"	48 "	12¼	Trumpet	
Butler Flats	41°36'14"	70°53'42"	Flg.W.	0m05s	"	53 "	12¾	Bell	
Palmer Island	41°37'3[illegible]"	70°54'35"	F.R.		"	34 "	8½	"	
Ned Point	41°39'08"	70°47'46"	"		White	40 "			
Bird Island	41°40'10"	70°43'04"	F.W.v.W.Fl.	1m20s	"	37 "	11¼	Bell	
Wings Neck	41°40'48"	70°39'42"	F.W.		"	50 "	12¼	"	
Hedge Fence (Lt. Ves.)			2 F.W.		Red	50 "	12½	Siren	Signals 4-1
Bishop and Clerks	41°34'28"	70°15'02"	Flg.W.R.Sec.	0m30s	Gray	56 "	13	Bell	

Fig. 44. Information in regard to lights and their various properties shown on a Government chart

vey was a good one, on a sandy bottom in a region where currents are strong, or the seas heavy, marked changes are liable to take place. When navigating by landmarks the chart of the locality which is on the larger scale should be used.

The depths of water are shown on the chart in various ways. On some larger scale charts they are recorded in feet at mean low water, and on others the depths on the clear portion of the chart are shown in fathoms, and those on the shady portion are given in feet, while on other charts all soundings are in fathoms. It is absolutely necessary to refer to the key on the particular chart in use to ascertain which method is followed in showing the depths.

As all depths are given for mean low water, it may be that the water is deeper or even shallower at the time of your sounding than that shown on the chart, and the proper correction should be made for the variable factors.

Different Kinds of Charts

Three Departments of the Government issue charts, as follows: The Coast & Geodetic Survey of the Department of Commerce publishes from its surveys charts which are suited to the purposes of navigation, commerce, and public defense. The Hydrographic Office in the Navy Department has charge of the duplication of charts and plans issued by other nations, and the publication of charts by the Navy of coasts not under the jurisdiction of the United States; the Corps of Engineers in the War Department issues charts of the Great Lakes.

There are four series of charts on the Atlantic, Gulf, Pacific, and Philippine Island Coasts, the first series consisting of sailing charts, which embrace long stretches of coasts—for instance, from the Bay of Fundy to Cape Hatteras. These are intended to serve for offshore navigation, or between distant points on the coast, as for example, Portland, Me., to Norfolk, Va. They are prepared for the use of the navigator in fixing his position as he approaches the coast from the open ocean, or when sailing between distant coast ports. They show the offshore soundings, the principal lights and outer buoys and landmarks visible at a great distance.

The second series is known as the general charts of the coast. They are on a scale three times as large as those of the first series, and embrace more limited areas, such as the Gulf of Maine, etc. They are intended for coastwise navigation when the vessel's course is mostly within sight of land, and her position can be fixed by landmarks, lights, buoys, and soundings.

The third series comprises the coast charts, which are constructed on a scale five times as large as that of the second series. One inch on these charts represents about one nautical mile, or one and one-seventh statute miles. They are intended for close coastwise navigation, for entering bays and harbors, and for navigating the large inland waterways.

The fourth series embraces the harbor charts, which are constructed on large scales intended to meet the needs of local navigation.

CHAPTER X

Publications and Nautical Instruments

IN order to navigate successfully any small motor craft along the thousands of miles of our sea coast, or in and around the numerous bays and harbors, it is not necessary for the motor boatman to provide himself with an expensive set of navigating instruments, such as is needed for deep sea navigation, but there are a few of the more simple instruments and other requisites, which should be aboard every craft, whether or not she goes out of sight of land. The cost of such instruments is trifling—in fact, many of them can be home made. These with a little care can be made to give as accurate results under ordinary conditions as the more expensive and refined instruments will.

Government Publications

Among the requisites necessary for this simple coastwise navigation, it is hardly necessary to enumerate the more common ones. Every motor boatman knows that he should have on board for successful piloting the best available chart of the locality to be traversed, together with the sailing directions and description of the aids to navigation, and, as was just pointed out in the last chapter, it is equally important that all of these should be brought up to date.

But the question is how to procure these sailing directions, charts, information about aids to navigation, etc. (See Fig. 45), and principally how to keep them up to date. Few motor boatmen realize that the various departments of the United States Government are doing all this work for them, and that most of this information can be had for the mere asking.

Keeping Charts Up to Date

Keeping the charts up to date is more or less of a serious and expensive question for the Government, and unfortunately a majority of the Government charts cannot be furnished gratis to everyone, but after the original chart has been purchased by the motor boatman at a price which is the actual cost to the Government of producing it, he is sure that the chart will last many years to come, and still be up to date if he cares to take the pains to keep it so.

Every week the Bureau of Lighthouses and the Coast & Geodetic Survey publish a pamphlet which shows in detail every change which has been made to any aid to navigation during that week, or which is proposed for the near future, together with much other valuable information, such as newly discovered rocks and shoals, and information in regard to shifting bars, new publications, lists of new editions of Government charts, together with a list of canceled editions, etc. This pamphlet, which is known as the Notice

Fig. 45. Government publications which should be aboard every motor boat

to Mariners, is distributed free of charge to any motor boatman who will apply for it to the Division of Publications, Department of Commerce, Washington, D. C.

The items in the Notice to Mariners are arranged in geographical order, starting with the eastern coast of Maine, continuing South to the Gulf Coast, then giving information in regard to the aids to navigation on the Great Lakes, Pacific Coast, Alaska, and the Philippine Islands. It clearly shows the numbers of Government charts on which the change occurs, the page numbers in the Light List, Buoy List and Coast Pilot as well. (See Fig. 46.) For example, the Notice to Mariners of Sept. 10 gave notice that the characteristics of the light of the canal approach gas and bell buoy located in Cape Cod Bay were to be changed on Sept. 24 to a flash every six seconds, but with no other change. It also stated that this change must be noted on Coast & Geodetic Survey Charts, No. 1,208, 1,107 and 1,000; that the

186. VIRGINIA—Chesapeake Bay—Main Channel—Old Point Comfort Light Station—Fog signal changed, February 10, to an electrically operated bell, to sound 1 stroke every 7½ *seconds*, and moved to the U. S. Engineer wharf, nearer the channel, and about 170 yards 212½° from the lighthouse. (No. 7, 1917.)

C. & G. Survey Charts 400, 1222, 77, 376, 1109.
Light List, Atlantic Coast, 1917, p. 160, No. 844.
Buoy List, 5th District, 1916, pp 17, 28.
Coast Pilot, Section C, 1916, p. 129.

Fig. 46. A paragraph from the Notice to Mariners

Light List for the Atlantic Coast edition of 1915 must be corrected on page 42, Light No. 182; that the Buoy List for the second District, edition for 1914, must be corrected on page 29, and that Part 3 of the Coast Pilot must be corrected on page 58. This notice is further arranged so that the corrections can be clipped out of the Notice to Mariners and pasted in the particular Light List in its proper position. All of this information is sent out every week by the Government free of charge to anyone who is interested enough to have his name added to the mailing list.

The Light List

The Department of Commerce publishes annually a Light List for the Atlantic and Gulf coasts which includes all lighted aids to navigation maintained by or under the authority of the United States Lighthouse Service. The Light List includes lighthouses, lighted beacons, lightvessels, lighted buoys and fog signals, but not unlighted buoys or beacons. In order to increase the convenience of the list to boatmen following or approaching the coast, the coast lights are printed in heavier type.

The Light List, which is also furnished free of charge, gives the name of each light and its character and period of light—that is, whether it is fixed, flashing, group-flashing, occulting, alternating, etc. It states the location of the light, for example "in six fathoms, off the northern point of Iron Bound Island," and further gives the latitude and longitude of the more important coast lights. The height of such light above the sea level, the miles it is visible and the candlepower are also included. A description of the structure, lightvessel, or buoy with the distance of the top of the lantern above the base is also given. The fog signal, whether it is a bell, horn or trumpet, and the number of strokes and blasts and the interval between them is valuable and necessary information for every motor boatman.

The Buoy List

The Lighthouse Service also publishes and supplies free of charge, separately for each lighthouse district, a Buoy List which gives a list of all buoys in that district, both lighted and unlighted, as well as all other aids to navigation. The Buoy List does not give as much and as important information as to the lights and fog signals as does the Light List. The Buoy List is published more for local use, and as far as the actual location of buoys is concerned, the larger scale charts are a much more certain source of information. Besides showing the location of the buoys clearly, most of the Government charts published by the Coast & Geodetic Survey contain all the necessary facts regarding the buoys. Moreover, the individual charts are corrected up to the date of their issue by the Government, while the Light and Buoy Lists can only be brought up to date by the Government when a new edition is published.

Coast Pilots

The importance of the motor boat has been recognized by the Government in the preparation of the new editions of Coast Pilots. Much additional data and information about points frequented by the motor boatman has been included in the new editions, which was decidedly lacking in the older publications. For example, the sailing directions are given for entering harbors, bays, inlets, etc., into which it is possible to carry only a few feet draft, and for places which particularly provide shelter and protection for small craft. Points where fuel, ice, and other supplies can be taken on by small boats are mentioned. In the Coast Pilots, aside from the sailing directions and detailed information for thousands of places on the coast, a great mass of general information useful to motor boatmen is given. The system of buoyage for the various districts is taken up; points where pilots and tow boats may be obtained; quarantine and bridge regulations; a list of dry docks and marine railways; a table giv[illegible] prevailing winds during each month of the year for a number of years; information about fog, when and where it is most likely to occur; points where storm warnings are displayed, and practical rules for determining the signs of an approaching storm; the best method to avoid storms; a list of United States Life Saving stations; the amount of variation of the compass at different points, etc.

Tide Tables

Other publications issued by the Department of Commerce which are of great use to the motor boatman are the Tide Tables. These are published annually in advance, and are in several different forms. One of them is the General Tide Table giving information about the tides at all of the more important ports of the world. Another edition in the Atlantic Coast Tide Table for eastern North America, and a third is the Pacific Coast Tide Tables for western North America, Eastern Asia and many island groups.

Besides giving the time of every high and low water for every day in the year, the Tide Tables give the height of each of these above mean low water. The tables for the Atlantic Coast contain these full predictions at twenty important stations on the Atlantic and Gulf Coasts, which are extended to about 1,100 subordinate stations by means of a table of tidal differences.

Tides and Currents

A careful distinction should be made between the vertical rise and fall of the tide, which is marked at the transition periods by a stationary height or stand. The tidal current is the horizontal transfer of water as the result of the difference in level, producing the flood and ebb, and the intermediate condition known as slack water. It seldom occurs that the turn of the tidal stream is exactly coincident with high and low water, and in some channels the current may outlast the vertical movement which produces it by as much as three hours. The effect of such a condition is that when the water is at a stand, the tidal stream is at its maximum, and when the current is slack, the rise or fall is going on with the greatest rapidity.

Generally speaking, the rise and fall and strength of current are at their minimum along straight stretches of coast upon the open ocean, while bays, inlets, and large rivers operate to augment the tidal effects, and it is in the vicinity of these that one finds the highest tides and strongest currents. The navigator need not be surprised in cruising along a coast to notice that his vessel is set more strongly toward or from the shore in passing an indentation, and that evidences of tide will appear more marked as he nears its mouth.

Time of High and Low Water

The prediction of tides for the tide tables is made by means of a remarkable machine, which was designed and constructed for the purpose in the office of the Coast & Geodetic Survey. A photographic reproduction of this machine is shown in Fig. 47. There are three different harmonic scales for varying changes of tide. To prepare the machine for predicting the tides at any port, the harmonic constants obtained from an analysis of the tidal observations for that port, together with certain astronomic data to adapt these constants to the particular year to be predicted, must be entered in the machine. The machine takes account of the harmonic constants for thirty-seven elements of the tide, and after it has been set with these constants, the turning of a crank moves a system of pointers over dials, from which may be read the height of the tide at any desired time, and also very readily the successive high and low waters for each day. The times and heights are directly tabulated by the operator into a form that is sent to the printer to serve as copy for the tide tables published by this Department. It requires from two to three hours for one person to set the machine with the constants for any station, and from

Fig. 47. The machine used by the Government to predict the time and heights of tides, at the various sea ports

seven to ten hours more to operate it, and tabulate the tides for an entire year. If this machine were operated for ten hours a day for 300 working days in a year, the operator could tabulate, ready for printing, the time and heights of every high and low water during the entire year for 270 different ports.

The Chip Log

The chip log is not used to any great extent at the present time, as it has been superseded by the patent log. The chip log (See Fig. 48) consists of three principal parts, known as the log-chip, log-line, and the log-glass. The log-chip is a thin piece of wood, weighted at one edge sufficiently to make it float upright in the water. As it is thrown overboard from the boat, it assumes a position at rest relative to the boat. In other words, the log-chip remains in a fixed position, and the boat sails away therefrom. A log-line is made fast to the log-chip in a suitable manner at one end, the other end of which is wound upon a reel kept aboard the boat. After the log-chip has been thrown overboard, the log-line will begin to unwind from the reel, which is generally mounted on a spindle to be held in one's hands. At a distance of 15 to 20 fathoms from the log-chip a permanent mark is placed to allow sufficient length for the log-chip to clear the wake of the boat. The rest of the line is divided

Fig. 48. The chip log

into lengths of 47 feet 3 inches, called "knots," by pieces of fish line thrust through the strands. The number of strands of line or knots which leave the reel or pass over the taffrail of the boat in a given time, in the time which it takes the log-glass to empty itself, will be an indication of the speed of the boat. The log-line is so attached to the chip that when the entire length of the line has been withdrawn from the reel the sharp pull will cause one of the attachments in the log-chip to disconnect, and the log-chip will then float in a position of least resistance, and can be drawn aboard the boat without difficulty.

The Patent Log

The log is a device for determining the distance which a boat has run through the water. There are three principal kinds of logs known as the patent log, chip log (page 77), and the ground log. Of these the patent log is the only one used to any extent to-day.

The patent log (Fig. 49) consists of a registering device, tow line, and a rotator. The registering device is generally made fast to some permanent position near the stern of the boat, and

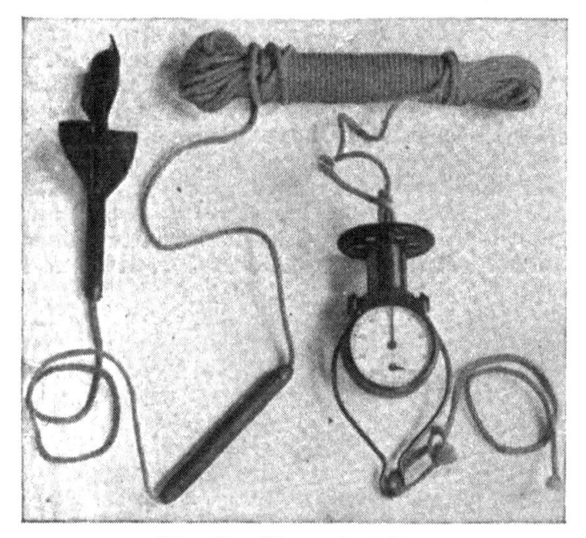

Fig. 49. The patent log

the length of tow line used must be sufficient to extend beyond the effects of the boat's wake. The rotator is attached to the outer end of the log line, and is a small spindle with a number of wings or blades extending radially in such a manner as to form a spiral, and when drawn through the water in the direction of its axis rotates about that axis after the manner of a screw propeller. The registering device is so calibrated that instead of registering the number of total turns of the rotator and log line, it will register in nautical miles.

The patent log is by no means perfect, especially when used on small boats. It is subject more or less to error, sometimes as great as 10 per cent. The patent log naturally registers distance which the boat has run in the water, and does not indicate distance relative to some point on land. In other

Fig. 50. The chronometer

words, to get the net actual distance run, allowance must be made in the reading of the patent log for currents, leeway and such errors. The patent log cannot be depended upon for actual readings at low speeds—that is, speeds of less than five or six miles per hour. Wave motion also has considerable effect upon the operation of the patent log.

The Ground Log

The ground log follows in a general way the principles which govern the chip log; except that a lead which sinks to the bottom is used in place of a chip, which floats in a fixed position. The ground log indicates speed relative to the land, and no corrections for current, etc., are necessary.

The Chronometer

The chronometer, the timepiece used on ships (See Fig. 50), differs from the clock or other timekeeping instruments in that it is constructed to perform its work with greater precision. Correct time, or regularity of a timepiece on shipboard is absolutely essential.

Chronometers should be handled with the greatest care and should not be subject to jarring, sudden shock or extreme changes in temperature.

As it is beyond the power of human skill to make anything absolutely perfect, it follows that all chronometers have more or less error. After this error has been reduced to a minimum there is no further effort made to adjust the chronometer. However, it is essential that the amount which the chronometer varies each day should be known, and the variation should be constant or nearly so. This error is known as the *rate* of the chronometer and the amount which the chronometer gains or loses each day is called the *daily rate*.

To insure a uniform rate the chronometer should be wound at the same time each day, although it may be capable of running a much longer period than twenty-four hours.

When using the chronometer for the purpose of observations, etc., a correction must be made to its reading for the chronometer's rate. For example, if the chronometer were known to be absolutely correct on June 1, and to have a daily rate of + 1/8 second, then when taking an observation on September 1, 11¼ seconds would have to be subtracted from the chronometer time to obtain the correct time.

The Lead Line

Of the instruments necessary for successful piloting, a motor boatman should be familiar with a few of the most common ones. Of these the lead line is probably the most useful. As is well known, the lead line is a device for determining the depth of water, and consists essentially of a suitably marked line having a piece of lead of a certain definite shape, somewhat similar to a window weight. For use on motor boats leads of various weights are used, ranging from five to fourteen pounds. A lead line of twenty-five fathoms is sufficient for all ordinary purposes. The deep sea lead weighs from 30 to 100 pounds, and a line of 100 fathoms or upwards is employed.

Marking the Lead Line

Lines are generally marked as follows:

2	fathoms from the lead, with	2 strips of leather.
3	" " " " "	3 " " "
5	" " " " "	a white rag.
7	" " " " "	" red "
10	" " " " "	leather having a hole in it.
13	" " " "	same as at 3 fathoms.
15	" " " "	" " " 5 "
17	" " " "	" " " 7 "
20	" " " "	with 2 knots.
25	" " " "	" 1 knot.
30	" " " "	" 3 knots.
35	" " " "	" 1 knot.
40	" " " "	" 4 knots.

And so on

Fathoms which correspond with the depths marked are called "marks." The intermediate fathoms are called "deeps." The only fractions of a fathom used are the half and quarter. The length of lead lines should be checked up frequently while wet. The bottom of the lead is hollowed out, and the hole is filled with tallow or a like substance by means of which a sample of the bottom is brought up. The process of filling the lead with tallow is called "arming the lead."

The Sounding Machine

The sounding machine, which replaces the deep sea lead and lead line, has several advantages over the latter, inasmuch as it permits of faster and more accurate soundings being made while the boat is moving through the water. It consists of a reel of strong wire mounted on a suitable stand and a controlling brake. Crank handles are provided for reeling in after the sounding has been made. A lead is attached to the outer end of the wire, above which is a cylindrical case containing the depth registering device.

Several forms of depth registering mechanisms are in use, one of which was devised by Lord Kelvin. In this, a slender glass tube is used, sealed at one end and open at the other. The inside of the tube is coated with a chemical substance which changes color upon contact with sea water. This tube is placed closed end up in the metal cylinder and as it sinks the water rises in the tube, the air being compressed by a force depending upon the depth of the water. The limit of discoloration is marked by a clear line and the corresponding depth is read off from a scale which goes with the sounding machine. With this pressure a slight correction is generally necessary to take account of the atmospheric pressure at the time the sounding is made.

Taking a Bearing

In order to take observations and bearings of distant objects with any degree of accuracy, it will be necessary to have some form of bearing finder or pelorus. The pelorus as it is manufactured and sold to-day by the dealers in nautical instruments is so very expensive that for the little use which the motor boatman has for such an instrument it would hardly pay him to go to the expense of purchasing one. However, with a little care, a home-made bearing finder may be constructed which under ordinary conditions will be found to give fairly accurate results.

Fig. 51 illustrates such a bearing finder. A circular piece of hard wood is cut out for the base of the instrument, and four series of tacks are placed 90 degrees apart, as shown. These tacks have the same function as the lubberline on the compass bowl. On the wood base a compass card is mounted; this may be one of those distributed for advertising purposes, or home-made, as the owner prefers. The compass card is mounted at its center to the base, so that it may be freely revolved around its center. Mounted at the same center will be seen a horizontal arm which rests directly on the compass card. This horizontal arm is provided at each end with vertical members, which are about six inches in height, and have small holes near their upper extremity, through which the distant object is sighted.

To use this bearing finder it is simply necessary to mount it in some suitable position on the boat, so that one set of tacks will represent the bow of the boat exactly as the lubberline on the compass does. The compass card of the bearing finder is then turned by hand, so that these tacks will be opposite that point on the compass card which represents the correct heading of the boat at the particular moment the bearing is to be taken. Keeping the pelorus base and its compass card in this position, the movable sighting vanes of the instrument are then turned so that it is possible to obtain a bearing of the desired distant object by sighting through the two holes in the uprights. (See Fig. 51.) The bearing of the object will then be indicated on the compass card below through an opening in the horizontal member of the sighting vanes. If possible, sights or bearings on a number of distant objects should be taken, and the bearings plotted on one's chart as a check.

If all the lines representing the bearings intersect in a common point, one may safely assume that the bearings are correct. It is hardly necessary to state that successive bearings should be taken as quickly as possible after one another, so that the boat has not covered any appreciable distance during the time.

Fig. 51. A home-made bearing finder, showing method of taking a bearing on a distant object

CHAPTER XI

Piloting

PILOTING, as the term is popularly known, is the art of conducting a vessel in channels and harbors along coasts where landmarks and aids to navigation are available for fixing the position, and where the depth of water and dangers to navigation are such as to require a constant watch to be kept upon a vessel's course, and frequent changes to be made therein.

Laying a Course

After one has become familiar with the different instruments used for piloting, including the dividers, parallel rulers, and course protractors, and if possible the chip log, patent log and the lead, he may consider himself qualified, following the information we have covered so far, to pilot a boat with safety. His first act, after providing himself with the best available chart of the locality to be traversed, together with the sailing directions and descriptions of the aids to navigation, all of which have been brought up to date as explained, will be to lay his course. This is done by marking one point upon the chart at the boat's position, and another point for which it is desired to steer. A line is then drawn connecting the two points, which will indicate the course to be steered by the boat. The motor boatman should examine carefully along this line on the chart to be sure that it clears all dangers.

Using the Course Protractor

The next step is to ascertain the magnetic direction of the line on the chart, representing the course to be steered. The course protractor has generally superseded the ancient parallel rulers for the purpose of transferring the direction of the line drawn on the chart to the compass rose on the chart, in order to determine its direction. In using the course protractor, its center should be placed on the chart exactly over the boat's position. The arm of the protractor is then swung around to the nearest compass rose on the chart, making the hair line down the center of the arm pass directly over the center of the compass rose. Holding the protractor arm firmly in this position, the compass part of the protractor is then swung around until the hair line cuts the same compass point or degree on the protractor compass as it cuts upon the compass rose. The compass and the rose are now parallel, or, in other words, they have the same variations. Holding the protractor compass firmly against the chart, the protractor arm is moved until its center line cuts the point on the chart where it is desired to lay a course. The compass course either in points or degrees can then be read off directly.

Should the boat's compass have any deviation, the course can be easily corrected to take account of this error by simply turning the protractor compass around, while holding the arm against the chart—clockwise if the error is easterly, and counter-clockwise if the error is westerly.

Locating One's Position

A navigator in sight of objects whose positions are shown upon the chart may locate his boat's position by several different methods. The choice of the method will be governed

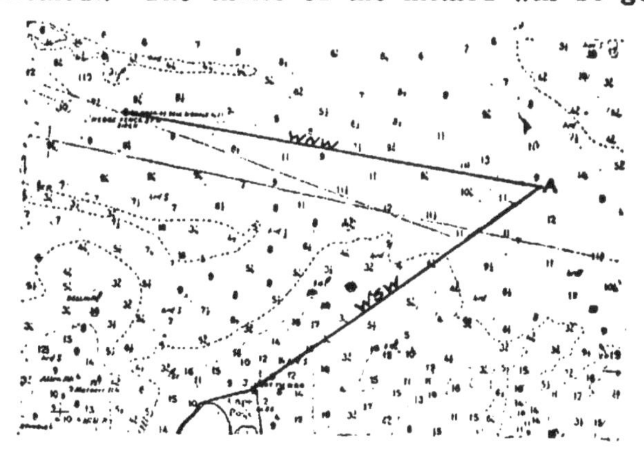

Fig. 52. Locating one's position by means of cross bearings

by circumstances depending upon which is best adapted to prevailing conditions.

Cross Bearings

One of the most frequently used methods of locating one's position is by means of what is known as cross bearings. Choose two objects whose position on the chart can be unmistakably identified, and whose respective bearings from the boat differ as nearly as possible by 90 degrees. (See Fig. 52.) Observe the bearing of each, either by compass or pelorus, taking one as quickly as possible after the other. See that the ship is on an even keel at the time the observation is made, and if a pelorus is used, be sure that the boat heads exactly on the course for which the pelorus is set. Correct the bearing so that they will be either true or magnetic according as they are to be plotted by the true or compass rose of the chart—that is, as observed by the compass apply deviation and variation to obtain the true bearing, or deviation alone to obtain the magnetic bearing. Draw on the chart by means of parallel rulers lines which pass through the respective objects in the direction that each was observed to bear. As the ship's position on the chart is known to be at some point on each of these lines, it must be at their intersection, the only point that fulfills both conditions.

If it be possible to avoid it, objects should not be selected for cross bearings which subtend an angle with a boat of less than 30 degrees, or more than 150 degrees, as in such a case a small error in an observed bearing gives a very large error in the result. For a similar reason objects near the ship should be taken in preference to those at a distance.

When a third object is available, the bearing of that may be taken and plotted. If this line intersects at the same point as the other two, the navigator may have a reasonable assurance that he has fixed his position correctly. If it does not, it indicates an error somewhere.

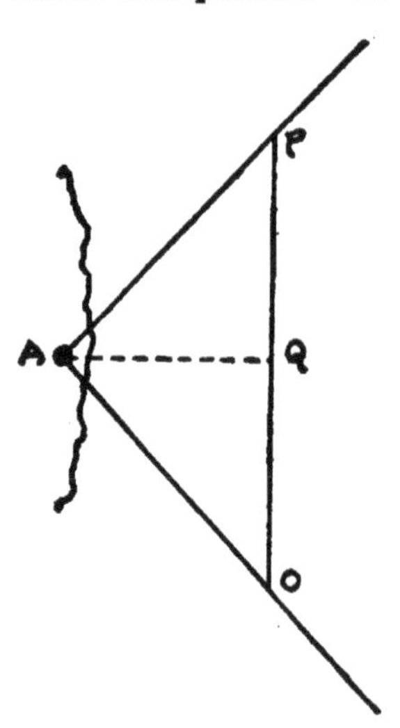

Fig. 53. Two bearings on an object

Two Bearings on an Object

Another method of locating one's position which is commonly used consists of taking two bearings on a known object. This is the most useful method, and certain special cases arise under it which is particularly easy of application. The process is to take a careful bearing of an object, and at the same moment read the patent log, then after running a convenient distance take a second bearing on the same object, and again read the log. The difference in readings gives the distance run.

In Fig. 53, line *OA*, represents the

direction of a known object *A* at the first observation; *PA* is the direction of the same object at the second observation, and *OP* is the distance run between the two observations. The problem is then to determine the point *P*, which locates the boat's position. This is accomplished by finding the distance *AP*, which is one side of the triangle *PAO*, and is done by referring to a well-known rule in trigonometry. It will be found $AP = OP \times \frac{\text{Sin POA}}{\text{Sin PAO}}$

Doubling the Angle

As has been said, there are certain cases of this problem which are exceptionally easy of application. These arise when the multiplier is equal to unity, and the distance run is, therefore, equal to the distance from the object. When the angular distance on the bow at the second bearing is twice as great as it was at the first bearing, the distance of the object from the ship at the second bearing is equal to the run.

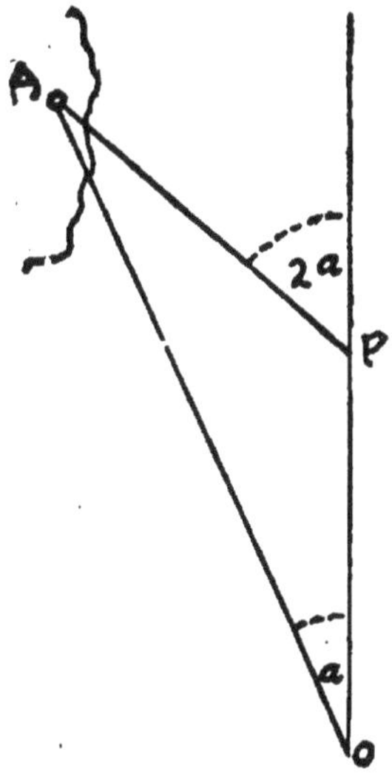

In Fig. 54 a bearing is taken of the object *A*, and this bearing is found to be equal to an angle which we shall call *a*. The boat is then held on the same course until the bearing of *A* has an angle twice the size of *a*. A boat is then at a distance *OP* from *A*, and as the direction *PA* is known, the point *P*, representing the boat's actual position, is easily found.

Fig. 54. Locating one's position by doubling the angle

Bow and Beam Bearing

A case where this method holds good is familiar to every navigator, and is known as the bow and beam bearing, where the first bearing is taken when the object is brought four points, or 45 degrees from ahead, and the second when the object is abeam. Then the distance run between observations will be identical.

In Fig. 55 the course of the boat is NE. The bearing is taken when the Gay Head Light appears East, which is at 45 degrees from the boat's course. The time is noted when this bearing is taken. The boat is then held on the same northeasterly course until Gay Head Light appears 90 degrees from the northeasterly course, or, in other words, bears SE. The time when the light bears SE is also noted. By knowing the speed of the boat in miles per hour one may readily calculate the distance from *A* to *B*, and the distance from *B* to Gay Head Light will be equal to *AB*, and thus the boat's position can be actually located.

26½° and 45° Bearings

Another case which is often made use of by mariners is known as the 26½-45 degree method. When the first bearing of an object is 26½ degrees from ahead, and the second bearing of the same object 45 degrees from ahead, then the distance at which the object will be passed abeam will be equal to the run between the two bearings. This method at once shows the navigator who is about to pass a point how wide a berth he is going to give the off-lying dangers.

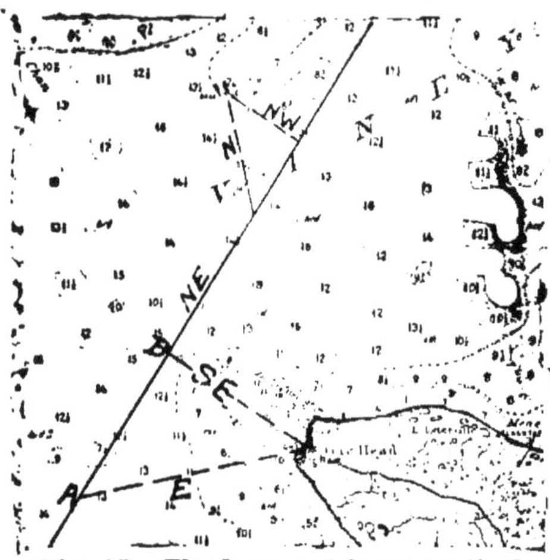

Fig. 55. The bow and beam method

In Fig. 56 the course is East. When the light on Bishop and Clerks Shoal bears 26½ degrees from the boat's easterly course, a note is made of the time. The boat is then held on a course until the same light bears 45 degrees, and the time again noted. The distance *AB* which is run between bearings can be calculated from the known speed of the boat. Then the distance at which the light will be passed, or the distance from the light to *C*, will equal *AB*. The mariner will at once know that he is passing far enough away from the light to clear Bishop and Clerks Shoal.

Distances on the water are deceptive and should not be relied upon in locating one's position except as a means of checking some other method. However, the motor boatman should practise determining distances on the water whenever possible. The charts give enough data as to the character of the shoreline, landmarks, heights of aids to navigation, etc., to be of great value in locating positions after one's eye becomes trained at determining distances.

Allowing for a Current

Piloting and navigating in waters where there is more or less current of a tidal or other nature is not quite as simple a proposition as it is in slack water. When courses are plotted, allowance must be made for the effect of a current depending upon whether the direction is such as to assist or retard the progress of the boat through the water.

The simplest case of the current problem is when the current is either directly ahead or fair. In such a case the resultant speed of the boat becomes simply the algebraic sum of the speed of the boat and that of the current. That is, if a current of 3 miles an hour is flowing in the direction in which the boat is sailing, the boat will have a speed of 3 miles an hour faster than she is capable of in slack or still water. If a 3-mile current is flowing in the opposite direction from that of the boat, the net speed of the boat will be 3 miles an hour less than her normal speed in still water. In other words, if the course happens to be for a distance of twenty-four nautical miles up a given river where there is a 4-knot current flowing down the river, and the normal speed of the boat in still water is 8 knots, then her net speed will be only 4 knots, and it will require six hours to cover the twenty-four-mile course. When the same boat comes down the river her speed relative to some point on shore will be 12 knots, and only two hours will be required to cover the distance.

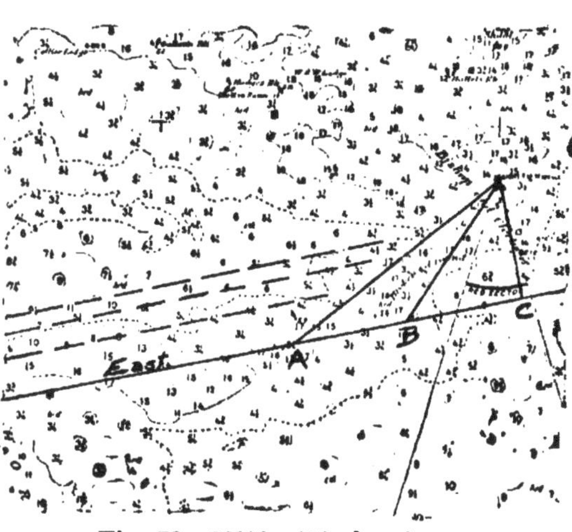

Fig. 56. 26½°—45°, bearings

With and Against a Current

It is generally assumed that should a boat sail up a river for a certain distance against the current, and then turn and come down the river for the same distance, her average speed in covering the whole distance will be the same as it would in slack water. However, such an assumption is incorrect, as the average speed of a boat under such conditions would be decidedly less than the speed she was capable of in still or slack water. For example see in Fig. 57 a course of eight nautical miles up a river which has a 4-knot current flowing down, and a boat with a normal speed in still water of 8 knots. The time required for the boat to go up stream—that is, eight miles against a 4-knot current—will be two hours, and the time to return down stream with a 4-knot current will be only two-thirds of an hour. Thus, the total time required to go the sixteen miles, one-half of which is with the current, and the other half against the current, will be two and two-third hours. Dividing this time into the distance (sixteen nautical miles), gives us a speed of 6 knots as the average for the entire trip. This amount is 2 knots slower than the normal speed of the boat in slack water.

S=Normal speed of boat=8 knots
C=Speed of current =4 knots

Then time required to go 8 (nautical) miles up stream=2 hours
And time to go 8 miles down stream = 2/3 hour
Total time to go 16 miles (½ with and ½ against current) =2-2/3 hours

$$A = \text{Average speed} = \frac{16}{2.67} = 6 \text{ knots}$$

$$\text{Time} = \frac{L \times S}{S^2 - C^2}$$

$$\text{Normal speed} = \frac{A + \sqrt{A^2 + 4C^2}}{2}$$

Fig. 57. Speed with and against the current

Across the Current

The next case in current navigation is when the current is directly across or at right angles to the boat's course. In Fig. 58 the boat's position is assumed to be at *A*, and the objective point it is desired to reach is *B*. The distance between *A* and *B* is taken as sixteen miles, and the direction north and south. Flowing at right angles to the course *AB* there is assumed to be a 2-knot current.

Under normal conditions it would require two hours for a boat having 8-knot speed to go from *A* to *B*. In crossing from *A* to *B* in two hours a boat would be carried down stream a distance of four nautical miles, inasmuch as there is a 2-knot current in this case, which would be operating for two hours. One will immediately see that should he attempt to steer in a southerly direction, he will, instead of reaching *B*, the desired destination, be carried down stream and will finally land at *D*, although his compass has indicated a southerly course all of the time.

To counteract the effect of the tide it is apparent that the boat must be headed up stream a trifle; that is, instead of holding a course due south, he must hold a course somewhat west of south.

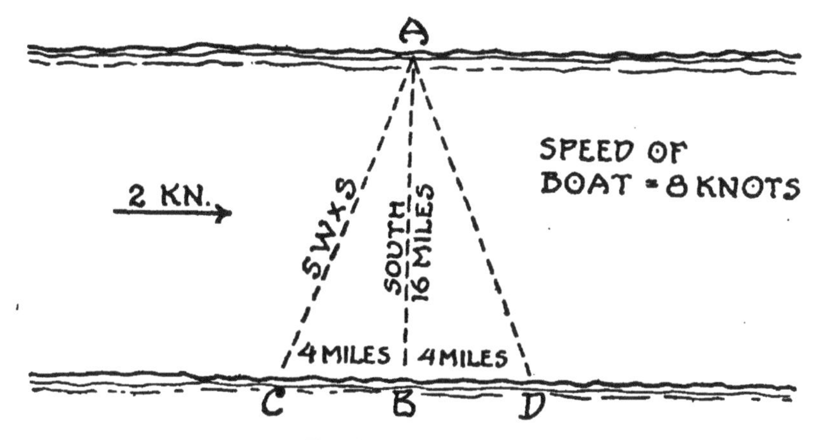

Fig. 58. Method of allowing for a current across one's course

To determine this course it is simply necessary to lay off on a chart to some convenient scale points *A* and *B* in a north and south direction from each other, and sixteen miles apart.

Then lay off from *B* at right angles to the line *AD* a distance equal to four miles, which, as explained above, is the amount which the boat would be carried away from *B* in the time it takes her to cross from *A* to *B*. This gives us a point four miles up stream, which may be called *C*. A line drawn from *A* to *C* will then be the direction which must be considered to counteract the effect of the current. By transferring the direction of this line *AC* to the compass rose on the chart by means of parallel rulers, or the course protractor, its direction may be found. In the case just cited the direction of the course will be found to be approximately SW by S.

Oblique Currents

If the direction of the current is other than directly ahead, astern or at right angles to one's course, the case is not quite as simple, but the course to be steered under conditions of a diagonal cross current may be quite readily explained by means of a diagram. (See Fig. 59.)

Suppose the boat is located at point *A* on the chart and the helmsman desires to lay a course to point *B*. The distance between *A* and *B* is eight miles, the direction North, the speed of the boat 10 miles per hour, and a tide setting NE at a rate of 2 miles an hour. What course should he steer and how long will it require to make the trip? Proceed as follows to some convenient scale say, one inch to the mile, and locate *A* and *B* in the north and south direction. Now from *A*, lay off a line opposite in direction to that of the current—that is, SW, and of a length equal to an hour's flow of the current—in this case 2 miles. From the far end of the current line draw a line *CX* parallel to the course line *AB*. Now find a point on *CX* which is of a distance from *A* equal to the normal hour's speed of the boat—in this case 10 miles—10 miles $= AD$. Draw a line from *A* to *D*, giving the direction to be steered. To find the time to hold this course in order to reach *B*, lay off a line from *B*, parallel to the current line *AC* which intersects the course line *AD* at *E*. Then a computation of the time it will take the boat in question, whose speed is 10 miles an hour, to go a distance equal to *AE*, or $7\frac{1}{8}$ miles, will give the time required to make the run from *A* to *B* with the current as stated.

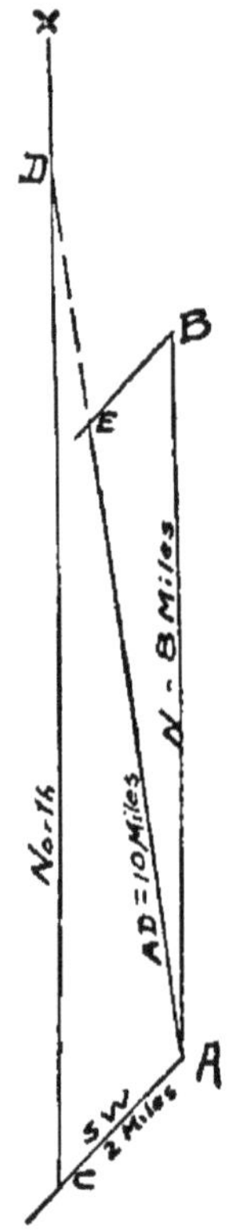

Fig. 59. Allowing for oblique currents

Use of Soundings

Soundings must not be regarded as definitely fixing one's position, but they afford a check upon other methods. Exact agreement with the chart may not and probably will not follow, as some inaccuracies may be expected, especially if the vessel is proceeding through the water. The height of the tide at the time the sounding is taken is not always a factor which can be determined with absolute accuracy. The soundings should agree in a general way with the chart, as should the nature of the bottom, and any very great departure should be cause for using great caution. This is especially true if the depth is found to be less than was expected. It is best to take soundings at regular intervals, knowing, of course, the speed at which the boat is going through the water.

By marking the soundings on a piece of tracing cloth, placing them the correct distance apart according to the scale of the chart being used at the time, along a line representing the track of the vessel, and then moving the paper over the chart, keeping the various courses parallel to the corresponding directions on the chart, until the observed soundings agree with those laid down, the ship's position will be well determined.

Identification

Before coming within the range of a light, a navigator should acquaint himself with its characteristics, so that when sighted it will be recognized. The charts, sailing directions and light lists give information as to the color, characteristic and range of visibility of the various lights. Care should be taken to note all of these, and compare them when the light is seen. If the light is of the flashing or occulting variety,

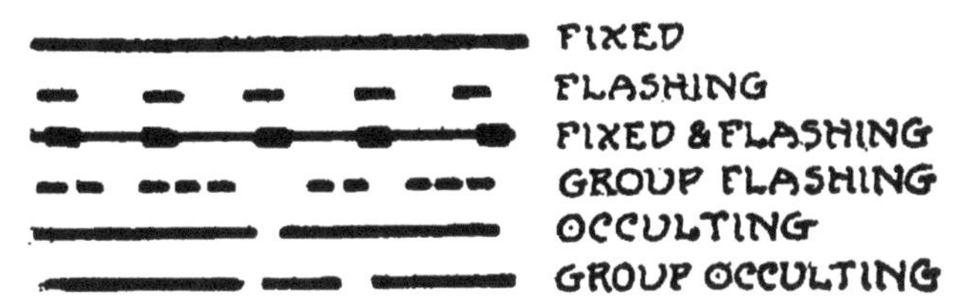

Fig. 60. Characteristics of navigation lights

the duration of its period should be noted. If a fixed light, a method that may be employed to make sure that it is not a vessel light, is to descend several feet immediately after sighting it, and observe if it disappears from view; a navigation light usually will do so excepting in misty weather, while a vessel's light will not. The reason for this is that the navigation lights are as a rule sufficiently powerful to be seen at the farthest point to which the ray can reach without being interrupted by the earth's curvature. They are, therefore, seen at the first moment that the ray reaches an observer on a ship's deck, and are cut off if he lowers the eye. A vessel's light, on the other hand, is usually limited by its intensity, and does not carry beyond a distance within which it is visible at all heights.

Peculiarities

In approaching a light of varying intensity, such as fixed, varied by flashes, or alternating white and red, due allowance must be made for the inferior brightness of the less powerful part of the light. The first named light may, on account of distance or haze, show flashes only, and the true characteristic will not be observed until the observer comes within the range of the fixed light. An alternating red and white light may show an occulting white until the observer comes within the range of the red light. Also where there are two fixed lights, one white and one red, the latter may be obscured, and the station may appear to show only a fixed white light.

Locating One's Position by the Bow and Beam Methods

Miles traveled between 1st bearing and 90° bearing.	Difference in Points Between Ship's Course and First Bearing.					
	2	3	4	5	6	7
1	.409	.667	1	1.5	2.414	5.028
2	.816	1.33	2	3.0	4.828	10.056
3	1.227	2.00	3	4.5	7.242	15.084
4	1.636	2.67	4	6.0	9.664	20.112
5	2.045	3.33	5	7.5	12.07	25.240
6	5.454	4.00	6	9.0	14.48	30.17
7	2.863	4.67	7	10.5	16.90	35.20
8	3.272	5.33	8	12.0	19.31	40.224
9	3.681	6.00	9	13.5	21.73	45.252
10	4.09	6.67	10	15.0	24.14	50.28
11	4.50	7.33	11	16.5	26.55	55.31
12	4.91	8.00	12	18.0	28.97	60.34

Directions: Take bearing of an object when it bears either 2, 3, 4, 5, 6, or 7 points over bow and note time. Hold course till some object is abeam and note time again. From the known speed of boat compute distance traveled between two bearings. Read down left hand column to figure corresponding to distance traveled between bearings and then following the line over to the right until you reach the column with the heading corresponding to the first bearing and read off distance you are away from object when it is abeam.

CHAPTER XII

Navigating in Fog

SOUND is very erratic over the water, and dependence upon it alone has cost many a life at sea and many a vessel. Often there are belts and areas over which sound does not seem to carry, and other phenomena also occur which change the direction that sound travels.

Precautions

Mariners are cautioned that, while every endeavor will be made to start fog signals as soon as possible after signs of fog have been observed, they should not, when approaching the land in a fog, rely implicitly upon these fog signals, but should always use the lead, which in most cases will give sufficient warning. A fog often creeps imperceptibly toward the land and a vessel may have been in it some time before it is observed at a lighthouse. As sound is conveyed irregularly through the atmosphere, mariners are strongly cautioned that they must not place dependence on judging their distance from a fog signal by the power of the sound. Under certain conditions of the atmosphere the sound may be lost a short distance from the station, as there may be silent areas or zones, or the sound may carry much farther in one direction than in another, and these conditions may vary in the same locality within short intervals of time. Mariners must never assume that the fog signal is not in operation because they do not hear it, even when they are in close proximity to it. The above applies particularly to fog signals sounded in air, as steam or air whistles, sirens, horns, or ordinary bells. Attention should be given to observing a fog signal in positions where the noises of the ship are least likely to interfere with the hearing, as experience shows that though such a signal may not be heard from the deck or bridge when the engines are running it may be heard when the ship is stopped or from a quiet position; it may sometimes be heard from aloft, though not on deck.

The rules prescribed for use in fog are also to be followed both day and night in mist, falling snow, or heavy rain storms.

The velocity of sound through the air depends to some extent upon the temperature. At 32° F., sound travels at the rate of 1,093 feet per second; at 62° F., 1,126 feet per second, and at 90° F., it has a speed of 1,155 per second.

Fog Signal for Power Vessel

Vessels falling in the class of steam vessels when under way should sound the fog signal on the whistle or siren, and sailing vessels and vessels towed should use the fog-horn.

A prolonged fog blast means one of from four to six seconds' duration.

The Inland Rules provide that a power vessel shall sound, when under way, one prolonged blast, at intervals of not more than one minute.

One of the principal rules of navigation in a fog is to use the greatest caution at all times, keeping the speed of the boat moderate, and having a careful regard to existing circumstances and conditions. When a fog signal of an approaching vessel is heard it is the duty of every captain to keep his vessel under absolute control, stopping if necessary until the danger of collision is over.

Sailing Vessel

A sailing vessel on the starboard tack sounds one blast of the fog-horn every minute and when on the port tack two blasts of the fog-horn in succession every minute, and when the wind is abaft the beam, three blasts in succession.

Boat at Anchor

Any vessel at anchor must ring the fog bell rapidly for a period of five seconds every minute.

Vessel Towing or Being Towed

Any vessel which is towing, being towed, engaged in work on a cable, or by accident or for any other reason cannot get out of the way of an approaching vessel must give a prolonged blast followed by two short blasts, at intervals not exceeding one minute.

Rafts

Rafts or other craft not specified shall sound a blast of the fog-horn or equivalent signal at least every minute.

Fishing Vessels

Fishing vessels, as drifters, trawlers, dredgers, and the line-fishing craft, if over 20 tons gross, must, when engaged in fishing, give a prolonged blast on the whistle or fog-horn according to whether they are driven by steam or sail, the blast to be immediately followed by ringing the bell.

Boat Aground

Any vessel which has the misfortune to run aground, or, in the case of a fishing vessel, gets her gear fast to a rock or other obstruction, is considered at anchor, and must make the signal necessary for such case.

Under Way with No Way On

A vessel under way, but with no way on, sounds two prolonged blasts in rapid succession every two minutes, according to the International Rules. The Inland Rules make no special provision for a boat under way but with no way on, and the regular fog signal for a boat with way on is generally used in this case.

Fog Stations

The first fog signal in the United States was a cannon, installed at Boston Light in 1719, which was fired when necessary to answer the signals of ships in thick weather. Bells were introduced at a comparatively early date, and at first were usually small, and rung by hand, to answer vessels. Trumpets were introduced in 1855. The original device consisted of a steel reed or tongue enclosed in a box with a large trumpet, the apparatus being sounded by means of compressed air produced by means of horse power. Steam whistles were first investigated in 1855, and were used for some time, but have been abandoned on account of the rapid deterioration of boilers, the expense of providing fresh water and fuel, the possibility of confusion with the whistle of a passing vessel, and above all, the time required to place this signal in operation in the event of a sudden fog.

Sirens were first employed in 1867, compressed air being used generally as a sounding medium. The compressors are now driven by internal combustion engines. Practically all fog signals as now installed are provided with a governing device for timing the strokes of blasts. In order to guard against the possibility of breakdown, all modern fog signaling installations are in duplicate, so that the second signal may be started at once in the event of accident to the first.

CHAPTER XIII

Flags and Colors

A YACHT or other pleasure boat is known by the flags she flies. There is nothing that gives a poorer impression to those who know than to observe the misuse of flags and colors. A visiting yacht is immediately sized up when she enters the anchorage of a club where she intends to spend the night, or a few days, by just these little things. The rules for flying flags and colors are very simple, and there is no excuse whatsoever for breaches of such rules.

Yachting etiquette, including the proper colors to fly, and their correct location, is largely governed by custom, but the fundamental rules in this respect were established many years ago before the advent of the motor boat. They are not in every instance well suited to the modern craft.

Time for Flying Colors

In general, flags are flown from 8 A. M. to sundown. There are a few exceptions to this rule regarding special flags which will be mentioned later. The time for colors should be taken from the boat of the senior officer present, except when one is in the vicinity of a U. S. naval vessel, or a naval station, in which case the boat of the senior officer takes the time for colors from the naval vessel or station. It is not permissible to fly more than one flag from the same hoist, nor a flag with a name spelled out thereon. This is a most terrible breach of etiquette for which there is no excuse.

Early and Late Colors

While it is a hard and fast rule that colors are made at 8 A. M., and hauled down at sunset, yet when a boat gets under way from a strange anchorage earlier than 8 A. M. she should immediately hoist her colors provided it is daylight, and she should keep them flying unless she comes to anchor before 8 A. M., when they should be hauled down. When a boat enters a strange anchorage, or gets under way after sunset, but during daylight, she should have her colors flying, but should haul them down before coming to anchor. It is also permissible to fly the yacht ensign on festive occasions after sundown while a yacht is illuminated for some special purpose.

Fig. 61. Various flags and colors flown on motor boats

Shapes Used for Flags and Colors

Three shapes are used in making up flags and colors—namely, triangular, swallow-tail and rectangular. (See Fig. 61.) Club flags are invariably triangular in shape. Owners' private signals are generally swallow-tail, and all other flags and colors are rectangular. National colors, as well as flag officers' flags, are rectangular. Of the latter three colors are used to distinguish the rank of the officers. Blue is used for the officer of the highest rank, generally commodore; red for the officer next in rank, generally vice-commodore, and white for rear commodore.

Special Flags

There are certain other special flags (See Fig. 61), such as the church pennant, flown above the yacht ensign, or national ensign at the stern, during divine service on board. The guest flag, a blue rectangular flag with a white diagonal bar across it, is flown from the starboard spreader from daylight to dark when the owner is absent and guests are on board. The meal flag, which is a white rectangular flag, is flown during meal hours of the owner in daylight from the starboard spreader when at anchor only. The crew's meal flag, which is red in color, and triangular in shape, is flown from the port spreader during meal hours of the crew when the boat is at anchor only. The absent flag, which is a blue rectangular flag, is flown from the starboard spreader from daylight to dark during the absence of the owner.

Flags for Open Boats

On boats of the type commonly referred to as open boats (See Fig. 62) having only bow and stern staffs, it is the

Fig. 62. Colors carried on an open boat

Fig. 63. Correct practice provides for flying the owner's private signal forward on a boat of this type when under way

common practice to fly the club flag from the bow staff, and the yacht ensign from the stern staff. However, correct practice (See Fig. 63) provides that this type of boat shall fly the owner's private signal at the bow staff, while the boat is under way, and the club signal when she is at anchor.

Boat with Signal Mast

A boat having bow and stern staffs, together with a signal mast amidships (See Fig. 66) flies the club signal from the bow staff, the owner's private signal from the mast-head at

Fig. 64. It is common practice to fly the club burgee forward, and the ensign aft on cruisers of this type

Fig. 65. Boats having a single mast and owned by a flag officer fly the club burgee forward, the officer's flag at the mast-head, and the yacht ensign aft

the signal mast, and the yacht ensign from the stern. If the owner of such a craft happens to be a flag officer of a club, be substitutes his officer's flag (See Fig. 65) for the private signal. Officers' flags are flown at all times while the boat is in commission, that is, during day and night, unless the owner is cruising with, or at an anchorage of a club of which he is a member, but not an officer. In such a case he should substitute his private signal for his officer's flag for the time being. Owners' private signals are flown only from 8 A. M. until sundown.

Flying the Jack

The Union Jack is flown from the bow staff when at anchor only, and only on Sundays and holidays, or on days of festive occasion. The Union Jack is never flown when the boat is under way under any circumstances. This is a rule which is very commonly broken, and, of course, does not apply to commercial craft.

Boats with Two Masts

Boats having bow and stern staffs, together with two masts (See Fig. 66), fly their club signal from the foremast head, their private signal or officer's flag from the mainmast head, and the yacht ensign from the stern. Nothing is flown from the bow staff while the boat is under way. When at anchor on Sundays or holidays, she flies the Union Jack from the bow staff.

Flags for Dinghies

A motor dinghy belonging to some larger craft should fly the yacht ensign from the stern staff when the dinghy is at the boat's gangway or away from the boat. From the bow staff the dinghy should fly the owner's private signal when he is aboard. When the owner is not aboard the dinghy, but there is some club member aboard, the club signal should be flown from the bow staff. When there is neither owner nor club member aboard, no flag should be flown from the bow. Flags on dinghies should not be flown when the dinghy is made fast astern or to the boat boom or is being towed. The flag of the senior person present takes precedence.

Half-Masted Colors

On occasions of national mourning, the ensign only should be half-masted. On the death of an owner, his private signal, and the club burgee on his boat should be half-masted, as

Fig. 66. On boats with two masts the club flag is flown at the fore, and the owner's private signal at the main. On boats with only one mast the club burgee is flown at the bow staff, and the owner's private signal at the mast-head

well as the club burgee on other members' boats, both while the boats are at anchor and under way.

In half-masting colors they should, if not previously hoisted, be mast-headed first, and then lowered to half mast. Before lowering colors from half mast, they should first be mast-headed, and then lowered. When an ensign is at half mast it should be mast-headed before making a salute by dipping the ensign.

What Club Flags to Fly

The owner should fly the club burgee of the club at which he is anchored, or whose fleet he is with, provided he is a member

Fig. 67. Day signal of a pilot vessel. A blue flag at the main mast-head

of that club; otherwise he may elect to fly the burgee of any club of which he is a member.

In making colors at 8 A. M. the ensign should always be raised before other flags and colors, and in hauling down at sunset the ensign should be the last flag lowered. Under no circumstance should the ensign be allowed to touch the water or the deck.

CHAPTER XIV

Yachting Etiquette

ALL salutes, whenever possible, should be made by dipping the ensign once. (See Fig. 66.) Whistles should never be used in saluting, and gun salutes should be avoided.

The salute for passing boats is one dip of the ensign. Guns should not be fired on Sunday or between sunset and sunrise for any reason whatsoever, except as signals of distress.

The gun for colors at 8 o'clock in the morning and at sunset in the evening should be fired from the yacht of the senior officer with the fleet, whether or not the officer is on board.

Upon entering an anchorage, captains should salute the commanding officer of the anchorage by firing one gun or by dipping the ensign once at the moment of letting go the anchor.

On ordinary occasions, when the commodore's yacht enters a harbor, his flag should be saluted by one gun or by dipping the ensign from the yacht of the senior officer present, and this salute should be acknowledged in kind by the commodore. However, should the commodore be entering harbor to assume personal command of his squadron he should be saluted when he drops anchor, by the firing of one gun, or the dipping of the ensign by each yacht in the squadron. This salute should be acknowledged by one gun from the flagship.

When a junior flag officer's yacht enters harbor, his flag should be saluted when he drops anchor by one gun or the dipping of the ensign from the yacht of the senior officer present, provided the latter is inferior in rank to the arriving flag officer; if not, the inferior arriving officer should salute the flag of the officer in command of the anchorage with one gun or a single dipping of the ensign when his yacht anchors.

A senior officer leaving harbor should indicate that he has transferred his command to the officer next below him in rank, by firing a gun or by dipping his ensign once upon getting under way.

When a flag officer makes an official visit between colors and sunset his flag should be flown in place of the club flag on the yacht while he is on board; upon leaving, one gun should be fired after he has entered his dinghy and shoved off. His flag should then be lowered.

A salute to another club is given by firing one gun and hoisting the signal of that club at the bow. After the salute has been returned in kind or a reasonable time for the return of such salute allowed, the club signal should be hauled down and the yacht's own club signal hoisted. In the absence of the signal of the club which is being saluted, the yacht's own club signal may be half-masted.

In meeting under way, yachts should salute by simply dipping the ensign once and never by use of the whistle.

Salutes between squadrons of different clubs, or from a single yacht to a squadron should be exchanged only by the commanding officer of the squadron.

Persons junior in rank should enter small boats and tenders before their seniors and leave after them. When a motor boat lands alongside a float seniors should disembark first.

International Life-Saving Signals

1. Upon discovery of a wreck by night, the life-saving force will burn a red pyrotechnic light or a red rocket to signify, "You are seen; assistance will be given as soon as possible."

2. A red flag waved on shore by day, or a red light, red rocket, or red Roman candle displayed by night, will signify, "Haul away."

3. A white flag waved on shore by day, or a white light slowly swung back and forth, or a white rocket or white Roman candle fired by night, will signify, "Slack away."

4. Two flags, a white and a red, waved at the same time on shore by day, or two lights, a white and a red, slowly swung at the same time, or a blue pyrotechnic light burned by night, will signify, "Do not attempt to land in your own boats; it is impossible."

5. A man on shore beckoning by day, or two torches burning near together by night, will signify, "This is the best place to land."

Status of the Yacht Ensign

The United States yacht ensign was authorized by an Act of Congress about 1849. It is required by law that the yacht ensign be flown on documented yachts of over 15 tons' burden as a signal to indicate that the boat flying it is a pleasure boat. Undoubtedly it never was the intention of the original law providing the yacht ensign that it should be used at a substitute for the American national ensign. However, custom has made it the only American flag now flown on yachts.

Undocumented boats of less than 15 tons' burden are not required by law to fly any particular ensign or national flag. Such boats may fly whatever ensign they see fit.

CHAPTER XV

Signaling

THE International Code of Signals consists of twenty-six flags (See Fig. 68)—one for each letter of the alphabet—and a code pennant.

One-flag signals, B, C, D, L, P, Q, S, hoisted singly have a special significance. The code flag over each indicates that they are signals of a general nature of frequent use. Signal flags hoisted singly after numeral-signal No. 1, refer to the numerical table, as do also two-flag signals with the code flag under them.

Two-flag signals without code flag are urgent and important signals; with the code flag over them they are latitude, longitude, time, barometer and thermometer signals.

Three-flag signals express points of the compass, money, weights and measures, and other signals required for communication.

Four-flag signals with a burgee (A or B) uppermost are geographical signals; with C uppermost they are spelling or vocabulary signals; with G uppermost they are the names of men-of-war; with a square flag uppermost they are names of merchant vessels and are not in the signal book.

How to Make a Signal

In the following instructions the ship making the signal is called *A*; the ship signaled to is called *B*.

1. Ship *A* wishing to make a signal hoists her ensign with the code flag under it.

2. If more than one vessel or signal station is in sight, and the signal is intended for a particular vessel or signal station, ship *A* should indicate which vessel or signal station she is addressing by making the distinguishing signal (i. e., the signal letters) of the vessel or station with which she desires to communicate.

3. If the distinguishing signal is not known, ship *A* should make use of one of the signals DI to DQ.

4. When ship *A* has been answered by the vessel she is addressing (see paragraph 9), she proceeds with the signal which she desires to make, first hauling down her code flag, if it is required for making the signal.

5. Signals should always be hoisted where they can best be seen, and not necessarily at the mast-head.

6. Each hoist should be kept flying until ship *B* hoists her answering pennant "Close-up." (See paragraph 4.)

7. When ship *A* has finished signaling she hauls down her ensign and her code flag if the latter has not already been hauled down. (See paragraph 4.)

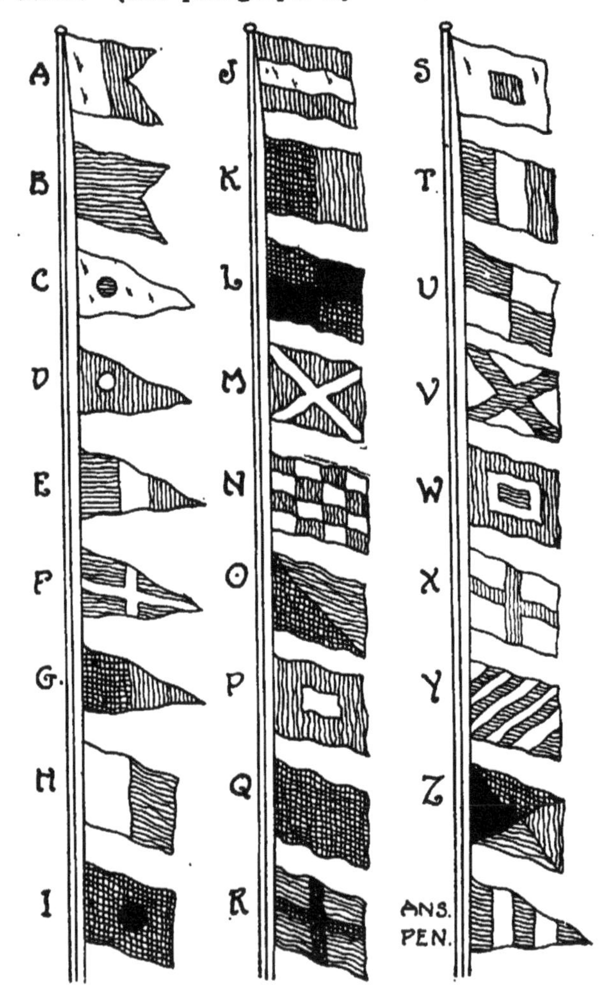

Fig. 68. The International Code flags (See page 99 for colors)

8. When it is desired to make a signal, it should be looked up in the general vocabulary, which is the index to the Signal Book.

How to Answer a Signal

9. Ship *B* (the ship signaled to) on seeing the signal made by ship *A*, hoists her answering pennant at the "dip."

(A flag is at the "dip" when it is hoisted about two-thirds of the way up—that is, some little distance below where it should be when hoisted "close-up."

The answering pennant should always be hoisted where it can best be seen.

10. When *A*'s hoist has been taken in, looked up in the Signal Book and is understood, *B* hoists her answering pennant "close-up" and keeps it there until *A* hauls her hoist down.

11. *B* then lowers her answering pennant to the "dip," and waits for the next hoist.

12. If the flags in *A*'s hoist cannot be made out, or if, when the flags are made out, the purport of the signal is not understood, *B* keeps her answering pennant at the "dip" and hoists the signal OWL or WCX, or such other signal as may meet the case, and when *A* has repeated or rectified her signal, and *B* thoroughly understands it, *B* hoists her answering pennant "close-up."

Semaphore Signal

Signals may also be transmitted by what is known as the two-arm semaphore method, using either hand flags or a machine for the purpose. The hand flags are from 12 to 15 inches square. They are blue, with a white square, or red and yellow diagonally. The one to be used depends upon the background. They are attached to a light wooden staff about two feet in length.

Fig. 69 shows the semaphore system of making signals by means of hand flags. This system is the most rapid method of sending spelled out messages. It is, however, very liable to error if the motions are slurred over or run together in an attempt to make speed.

Fig. 70 illustrates the semaphore machine which sends the signals in exactly the same way as indicated by the previous illustration, in which hand flags were substituted for the arms of the machine. The positions of the two methods are exactly the same. Both arms are moved rapidly and symmetrically, but there should be a perceptible pause at the end of each letter before making the motions for the next letter, and care must be

THE SEMAPHORE SYSTEM.

Characters	HAND FLAGS	Secondary Meanings	Characters	HAND FLAGS	Secondary Meanings	Characters	HAND FLAGS	Secondary Meanings	Characters	HAND FLAGS	Secondary Meanings
A		Error	I			Q			X		
B			J			R		Acknowledge	Y		
C		Repeat	K		Negative	S			Z		
D			L		Preparatory	T			Cornet		
E			M			U			Letters (follow)		
F			N		Annulling						
G			O		Interrogatory	V			Signals (follow)		
H		Execute	P		Affirmative	W			Interval		Designater

Fig. 69. Semaphore system of making signal by means of hand flags

taken with the hand flags to hold the staffs so as to form a prolongation of the arms.

Wigwag and Blinker Light Signaling

The dot and dash code (Fig. 71), comprising alphabets and numerals of the International Morse Code, is another method of signaling, frequently used on shipboard. In this system, which is known as the wigwag, messages are spelled out. There is one position with three motions. "P ition" is with the flag held vertically, the signal hand facing squarely towards the station with which it is desired to communicate. In the first motion (dot) the flag is waved to the right of the sender, embracing

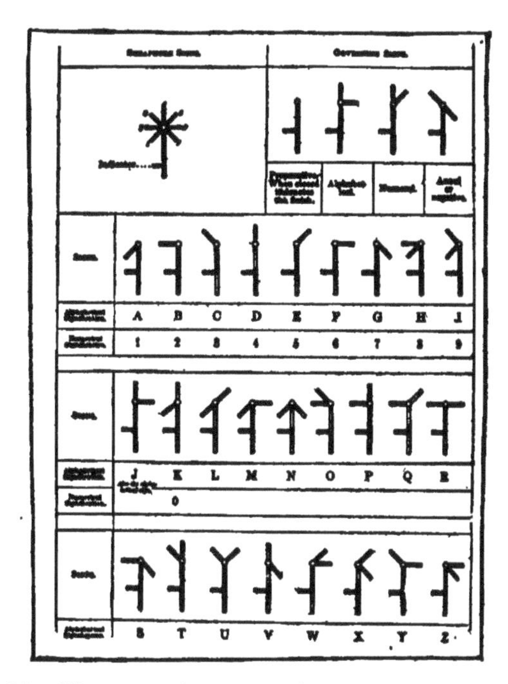

Fig. 70. The semaphore machine system of signaling

an arc of 90 degrees, starting from the vertical, and returning to it, and made in a plane at right angles to the line connecting the two stations. The second motion (dash) is a similar motion to the left of the sender. To make the third motion, "Front," which is used for separating words, sentences, etc., the flag is waved downward directly in front of the sender, and instantly returned to "Position."

It is important to obtain a good background, and to select a color of flag which will afford the most marked contrast with the background; to this end the red or the white flag shall be used as found best from the standpoint of visibility. When signaling to a considerable distance with a hand torch, electric portable, or hand lantern, a foot-light should be used as a point of reference to the motions; otherwise the methods are the same as for the hand flag. With an oil hand lantern a variation is permissible, as the lantern is more conveniently swung out and upward by hand from the footlight for "dot" and "dash" and raised vertically for "front."

To call a station, face it and make its call; if necessary to attract attention, wave the flag (or torch), making the call at frequent intervals. The station called makes "acknowledgment"; the sending station then makes "acknowledgment" and proceeds with the message. At night each boat called shall acknowledge by making her own call letter; the calling boat then makes her own call letter, which the receiving boats repeat; the calling boat then makes acknowledgment and proceeds with the message.

A day wigwag message for the entire force or for a group, flotilla, squadron division, or boat is indicated by the display, half yardarm high, of the cornet of the proper call. This is acknowledged by the boat or boats called hoisting the answering pennant half way; when all boats have thus answered, the message is proceeded with. At the end of a message sent as prescribed, the flagship hoists the call or the cornet, as the case may be, close up to the yardarm,

A · —
B — · · ·
C — · — ·
D — · ·
E ·
F · · — ·
G — — ·
H · · · ·
I · ·
J · — — —
K — · —
L · — · ·
M — —
N — ·
O — — —
P · — — ·
Q — — · —
R · — ·
S · · ·
T —
U · · —
V · · · —
W · — —
X — · · —
Y — · — —
Z — — · ·

Fig. 71. The International Morse Code for wig wag and wireless

whereupon, if the message is understood, the receiving ship or ships run the answering pennant close up to the yardarm. The hauling down of the call or cornet by the flagship is the acknowledgment of the answers, and the receiving boat or boats then haul down their answering pennants.

If, in the course of a signal, the sender discovers that he has made an error, he should make the characters corresponding to AA "Front," after which he proceeds with the signal, be-

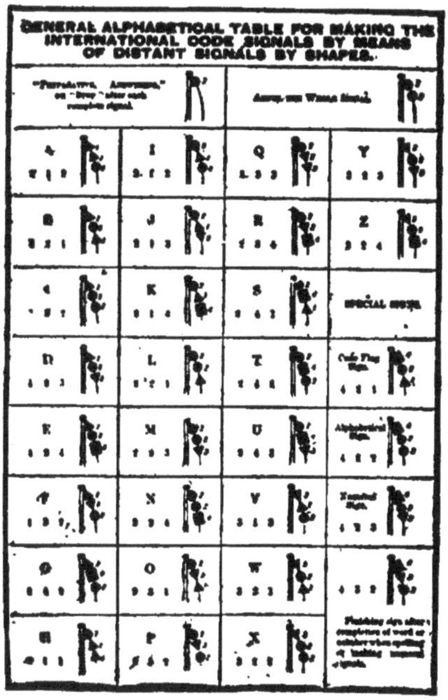

Fig. 72. The distance signals

ginning with the word in which the error occurred. If, in the course of a signal addressed to a single boat, the receiver does not understand a word, character, or display, he should "break in" with the characters corresponding to "repeat last word"; or, to have a whole message repeated he should make the displays which signify "repeat last message." In the case of a message addressed to several boats, an individual boat failing to understand a word shall not break in, but shall continue to read as much of the message as possible, and after the whole message has been sent shall request the next boat, or the division commander, or the commander in chief to repeat the missing portion.

Distance Signals

When in consequence of distance or atmospheric conditions it is impossible to distinguish the color or flags of the International Code, there is provided an alternative method of signaling, known as distance signals. (See Fig. 72.) There are three methods of making distance signals, as follows:

1. By cones, balls and drums.
2. By balls, square flags, pennants, and wafts.
3. By the Fixed Coast Semaphore.

In calm weather, or when the wind is blowing from and towards the observer, it is very difficult to distinguish with certainty any signal which depends on color or flags. The flags when used with shapes are also apt in calm weather to hide one of the balls or other shapes which would prevent the signal from being understood. Therefore, the system of cones, balls and drums is preferable to that of flags, pennants and wafts.

CHAPTER XVI

Miscellaneous Signals

THESE signals (Fig. 73) indicate the weather forecasts for twenty-four hours commencing at 8 A. M.

When displayed on poles the signal should be arranged to read downward; when displayed from horizontal supports, a small streamer should be attached to indicate the point from which the signals are to be read.

The morning forecasts (i. e., those issued from the A. M. reports) only, are utilized for the display of weather signals,

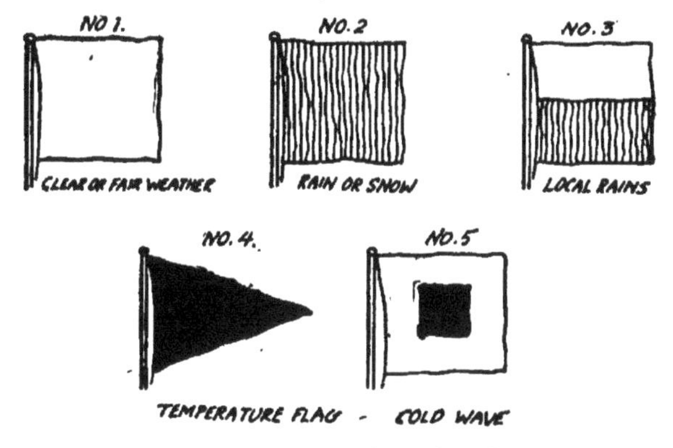

Fig. 73. Weather signals

and the flags displayed represent only the forecast applicable to the twenty-four hours beginning at 8 P. M. of the day the flags are hoisted.

If more than one kind of weather is predicted for the period from 8 P. M. to 8 P. M., the conditions first named in the forecast are represented by the uppermost weather flag in a vertical hoist, or by the weather flag nearest to the small streamer indicating the point, in a horizontal hoist, from which the signals are to be read. If two temperature forecasts are made for this period, the first-named only is represented by the temperature flag in its proper position. When the regular forecast contains warnings of a cold wave, the cold-wave signal is displayed alone.

and flags representing the weather element are never displayed on the same staff with the cold-wave signal.

If the forecasts contain a prediction, "moderate cold wave," "decidedly lower temperature," "much colder," etc., the cold-wave flag is not displayed, but the temperature flag is hoisted below the proper weather flag.

Flags are invariably lowered at sunset of the day the hoist is made, and no flags are displayed on the following day until the receipt of the next succeeding morning forecast.

The weather signal flags used are as follows (See Fig. 73):

Flag No. 1. Square white flag.
Flag No. 2. Square blue flag.
Flag No. 3. Square flag, white on upper half, and blue on lower half.
Flag No. 4. Triangular black flag.
Flag No. 5. Square white flag with black square in center.

Number 1 indicates clear or fair weather. Number 2 indicates rain or snow. Number 3 indicates that local rains or showers will occur, and that the rainfall will not be general. Number 4 always refers to temperature; when placed above numbers 1, 2 or 3, it indicates warmer weather; when placed below numbers 1, 2 or 3, it indicates colder weather; when not displayed, the indications are that the temperature will remain stationary, or that the change in temperature will not vary more than four degrees from the temperature of the same hour of the preceding day from March to October, inclusive, and not more than six degrees for the remaining months of the year. Number 5 indicates the approach of a sudden and decided fall in temperature. When number 5 is displayed, number 4 is always omitted. Examples:

Nos. 1 and 4, "Fair weather. Colder."
Nos. 4 and 2, "Warmer, Rain or Snow."
Nos. 4, 1, and 2, "Warmer, fair weather, followed by rain or snow."
Nos. 1 and 5, "Fair weather. Cold wave."

Distress Signals—In the Daytime

1. A gun or other explosive signal fired at intervals of about a minute.
2. The international code signal of distress, NC.
3. The distance signal consisting of a square flag, having either above or below it a ball, or something resembling a ball.
4. The continuous sounding of any fog-signaling apparatus.
5. The national ensign, hoisted upside down.

Distress Signals—At Night

1. A gun or other explosive signal fired at intervals of about a minute.
2. Flames on the vessel (as from a burning tar barrel, oil barrel, etc.).
3. Rockets or shells, throwing stars of any color, etc., fired one at a time at short intervals.
4. The continuous sounding of any fog-signaling apparatus.

Signals for a Pilot

A pilot may be obtained by displaying any of the following signals:

1. The International Code pilot signal indicated by PT.
2. The International Code flag S, with or without the code pennant over it.
3. The distance signal consisting of a cone, point upward having above it two balls or shapes resembling balls.
4. The Jack, hoisted at the fore.

At night—1. A blue pyrotechnic light burned every fifteen minutes.

2. A bright white light flashed at frequent intervals just a little above the deck.

To signal for a towboat set the ensign in the main rigging above the bulwarks for about a minute at a time.

Storm Signals

The warnings adopted by the United States Weather Bureau for announcing the approach of wind storms are as follows: (See Fig. 74.)

The storm warning (a red flag, eight feet square, with black center, three feet square), indicates that a storm of marked violence is expected. This flag is never used alone.

A red pennant (eight feet hoist and fifteen feet fly), displayed with the flags, indicates easterly winds, that is, from the northeast to south, inclusive, and that the storm center is approaching.

A white pennant (eight feet hoist and fifteen feet fly), displayed with the flags, indicates westerly winds, that is, from north to southwest, inclusive, and that the center has passed.

If the red pennant is hoisted above the storm warning, winds are expected from the northeast quadrant; when below, from the southeast quadrant.

If the white pennant is hoisted above the storm warning, winds are expected from the northwest quadrant; when below, from the southwest quadrant.

Night storm warnings—By night a red light will indicate easterly winds; a white above red light will indicate westerly winds.

The hurricane warning (two storm warning flags, red with black centers, displayed one above the other) indicates the expected approach of a tropical hurricane or of an extremely severe and dangerous storm.

No hurricane warnings are displayed at night.

A yellow flag with white center is a precautionary signal.

Signals should be read from the top of the staff downward.

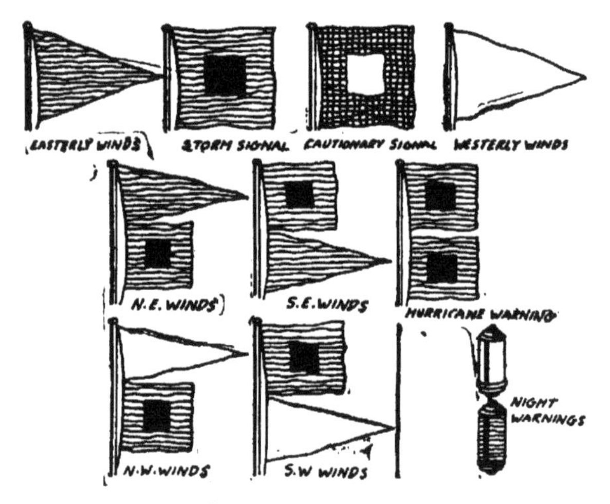

Fig. 74. The storm and wind signals

CHAPTER XVII

Boat Handling Under Various Sea Conditions

(*By E. K. Roden*)

WITH the advent of the motor boat as an important acquisition to our national defense, motor boating can no longer be looked upon merely as an exhilarative pastime. Its affiliation with military and naval forces guarding our extensive coast lines has placed motor boating upon a basis where responsibility and serious work go first, with sport and pleasure as a secondary consideration. This being the case, it becomes the duty of owners and operators of motor craft enrolled for defense service to acquire and cultivate not only a thorough understanding of the handling of the boat and engine, but also a knowledge of various wrinkles which in time of actual service may prove of value. The following suggestions apply in particular to a motor unit assigned to scout duty which through unforeseen exigencies may find itself in strange waters and, therefore, will have to rely upon the skill and resourcefulness of its commander to carry out his scout work and at the same time avoid the pitfalls likely to be encountered in navigating unknown sections of the coast.

Indications of Shallow Water

Upon approaching a shoal spot in the water, the attention of an observer will be attracted either by a rise in the height of the waves, with a tendency to curve over and break, or by their taking on a troubled, agitated appearance, in marked contrast to the waves in deeper water. The extent of the irregular water will, in most cases, clearly define the limits of the shoal, and, when traversing shoal places, the deepest water will always be found where the waves are of normal size and most regular appearance. They will be clearly distinguished from either the lifting kind, which are inclined to topple and break, or the smaller jumbled type. At times, the water over the shoal will be smooth and the water in the channels ruffled; this is particularly likely to be the case when the shoal bordering the channel has a growth of weeds reaching nearly to the surface.

If in strange waters and a line of ripples stretches across the course, the ripples should be approached with caution. The line may be caused by the changing of the tide, or it may be a reef or bar fairly close below the surface. These small ripples are often seen along the edge of shoals when the surrounding water is smooth, particularly when the outside water is deep; they are caused by the flow of the tide being shunted off by the shoal. During a strong breeze, when traversing a shoal having from 3 to 20 feet depth of water over it, the deeper parts may invariably be distinguished by watching for the heavier, more regular waves, while the shallow spots of the shoal are indicated by choppy, breaking waves.

Crossing a Bar

Necessity may at times compel the passage of a boat through a reef or a bar, over which a strong sea is running. In such cases it is well to run slowly along the reef at a moderate distance and search carefully for regular waves. If there is an opening, or channel, through, it will show water distinctly different from that over the rest. In such deep places, the water will remain without breaking until the sea has attained such violence that even the deep places have practically become shoals. The passage through comparatively unknown reefs and bars when heavy weather prevails should not be attempted except by the most experienced men. The sea may look smooth and regular at some distance off the bar, but on approaching, the conditions may be such as to require an intuitive skill at the helm to get the boat safely through.

Wave Motion

To run smoothly, a wave requires a depth of water as great as is the distance from its own trougn to trough. If that distance is 15 feet, the wave requires 15 feet of water to roll in or it will begin to rise in height and form a crest, this being the result of the friction of the wave motion on the bottom. It is the wave motion that travels, not the water, as can be readily seen by dropping a colored liquid of any kind into the sea—the color will remain stationary, or nearly so, while the motion of the wave will continue to advance.

When running along a beach at night, the beach being free from rocks, the line of safety can be felt by the lifting of the boat; if too close in, a sharp lift will be felt when a sea passes under—the motion being distinctly different from that felt when the boat was in deep water—and is a sure indication that the boat is within the line where the wave begins to top the breaker. In a heavy onshore wind, the best traveling will be found a mile or more offshore. The reason is that heavy seas on striking a beach or a reef give a strong recoil that causes a series of opposing waves which, meeting those coming in, produce rough, irregular water.

Occasionally there will be seen a solitary lift or leap of the water where there are no other evidences of disturbances; this is generally caused by a small mound or boulder arising at that spot from the bottom.

Tide Rips

Tide rips are the result of strong currents. With no visible signs of disturbance and the sea smooth all about them waves of this character will rear and tumble. They are clearly distinct from anything about them, and do not take one unawares. Almost invariably they have white foaming crests and roar in an unmistakable manner. Even in a white-cap breeze, they are clearly whiter than anything about them, and are so definitely marked that one can sail down their edges and admire the wildness of the scene. The wave motion in them is short and steep. When wind increases their turbulence, none but the staunchest of boats and best of helmsmen should attempt to enter the turmoil. When compelled to encounter them in bad weather the boat should be kept to the edges, where the water is always deep. If in the rip and it is running strong (which is generally the case during four hours out of the six), the boat should be kept head-to; she will lift and pound badly, and perhaps get strained, but that is better than the risk of rolling over. At the slack of the tide the rips do not exist.

Head Sea

During a hard blow, the sea will be found to present waves that are regular in general, but interspersed with seas that are too sharp for comfort for a boat of light construction. If going to windward, many of these seas will compel one to head into them; then will come a lift and, if the boat has not a sharp V-section forward, a smashing fall down the back of the wave. These falls pound the bottom of a boat so severely that it is not good practice to permit many of them. They are avoided by turning the boat a trifle off the wave, though if the swell is steep enough to throw the boat there is no help but to take it head on. Should swinging her off, to give more bearing surface, take the boat too much off the course, the remedy is to take the seas on one bow for a stated time and then on the other for the same interval, the result being an equalization of the course.

Following Sea

In running with a following sea the helmsman meets his hardest task. When a sea passes under the boat, lifting the bow, the next wave comes under the stern and begins to lift, and when the stern has been raised to a greater height than the bow, the latter starts to root—which means that the boat is "down by the head," and does not respond to the rudder. This is the anxious moment for the helmsman, as he waits for the feeling coming with a submerged rudder that indicates the direction in which the bow is going to turn. So long as the rudder is out of water it should be kept steady by the wheel, and the instant the feeling comes that it is submerged the helm should be turned, gently at first, then with all the strength necessary to counteract the sheer; then the wheel should be allowed to turn back freely as the boat balances on the forward drive on the face of the wave.

During the maneuvering the bow of the boat may root until two or three feet of it is buried in the sea ahead. It will not do to let this take place, for, as explained previously, the water is not moving, and the boat is plowing into it, and while doing so the stern may be lifted so high that she is deprived of her bearing, and will either dive or roll over. This is what occurs with boats trying to enter a surf or, when the weather is heavy from the seawards, to gain access to a harbor having a bar before the entrance. The remedy is the same in either case, and is the one commonly employed by life savers in making a landing; that is to tow a drag or sea anchor. If without one, any bulky article attached to a stout line may be dropped over the stern and towed. The resistance offered will help materially in checking the tendency to root.

With some boats and in some seas it will be found that the bow is rooting and the stern being boarded by the following waves. This is a bad case. All the movable weight (passengers, for instance) must then be placed amidships to lighten the ends and the bow swung a very little from a fore-and-aft bearing on the seas. If the boat is of the open-cockpit type, canvas should be fastened over the after end of the cockpit. This is a case in which oil might help some: the boat is run slowly and the oil put out from any part of the boat that will cause the slick to be spread by the time it reaches the stern. The burying of the bow and stern of a boat with fine sharp ends is of little consequence, as the lack of bearing surface in such a boat makes this a condition to be expected, but the sea has not hold enough on either end to do harm, and the end will rise as quickly as the wave passes by.

Beam Sea

In a beam sea, conditions are such as to require the utmost attention on the part of the helmsman. The boat is traveling in the trough and if an oncoming sea is a bad one, one must decide instantly whether to run or head into it. The present position of the boat generally governs the maneuver. If the boat has just recovered from a lurch and the bow is too far to windward to give her time to run off, she must of necessity be thrown head in. If she is too far off the wind to give her time to be swung up she must be sent to leeward. Most of the time the shape of the seas is such that the boat can be held to the course; this gives the helmsman the choice of the maneuver.

Lee Shore

When running along a lee shore for any considerable distance the scend of the sea will steadily set the boat toward the beach. There is seldom a sea so heavy that there are no smooth, well-rounded waves mingled with the rough ones, and in every smooth the boat should be sent on the course as far as she will go. Turn her to windward in the rough seas, and in some of the smooth ones if necessary, but in no case let her fall to leeward.

During a blow a boat should pass to the lee side of islands and shoals where it is posssible to do so; no shoal is so deep that it has no influence in smoothing the sea. A shoal near the surface will stop the waves altogether and leave only the wind for the boat to contend with.

Fog

If caught in a fog without a compass or with the compass out of order, the best way in which to prevent a boat from losing her direction is to take guidance from the run of the waves. Thus, if the waves were coming toward the starboard bow when the fog set in, they should be kept coming from the same quarter.

By trailing a line over the stern one may keep running straight ahead, and not in a circle as is often done. The longer the line the better, as with one of good length, any swerving from a straight course will show at once. Verification of the steering in a fog or rain may be gained by watching the slant of rain drops or drizzle.

Reflections of Rock and Sand

A majority of the rocks and shoals within range of the cruising motor boat are usually unmarked by buoys of any kind, but most of such obstructions betray their presence by reflecting their colors to the surface of the waters immediately surrounding them. The shade, or density, of the color will vary with the different phases of the day, from clear distinctiveness to an indefinable something, yet to the practised eye the hues may be distinguished and used with advantage. It well repays the operator of a motor boat to cultivate the faculty of observing the different shades of the water, as it gives a confidence in running that adds to the comfort and interest, and in combination with a judicious use of the lead line, enables him to pick his wav with a degree of certainty into harbors and inlets that are new to him. This applies more particularly to fairly clear waters and not such as are found in or close to the harbors of large cities. A mud bottom is not as good as a sand or rock bottom, but even over mud there will be different shades of color in the shallows and the channels.

When running in open waters, a faint line may appear at some distance ahead and commence to loom. On a near approach the entrance to a small harbor or inlet may be looked for, though the coast at first appears to be one unbroken line. As the boat draws nearer dark spots of brown may be seen at some places, while at others grayish or white shadows prevail; the former indicate deep and the latter shallow water. When approaching to four or five hundred feet, close observation will possibly show water of a decided greenish tint and water having a certain placid or slick whitish appearance. The latter color should be avoided and the deeper green followed, and then with a good lookout in the bow it will be perfectly safe to proceed slowly into the place as far as it is desired. The higher up the lookout is placed, the better he will see the bottom and select the route to be taken.

The entrance to an all-sand harbor over a bar may be made by observing the difference in color when arriving at the 18-foot depth, for the break is clearly visible. When passing into the 12-foot depth, it will be well to slow down to half speed. Here the darker green veins of water should be chosen.

FROM TO TOWARD DATE BAR. 8 AM NOON 8 P.M.

TIME	ABEAM OF	COURSE TO NEXT OBSERV.		DIST. BY CHART		SHOULD BE	PAT. LOG	SPEED		MOTOR	WIND		WEATHER	SEA	SOUNDINGS		
		MAGN.	COMPASS	TO NEXT OBSERV.	TOTAL FROM START	ABEAM OF NEXT OBSERV. AT	READING	SINCE LAST OBSERV.	AVERAGE SINCE START	R.P.M.	DIR.	FORCE			DEP.	BOT.	

FUEL	HIGH WATER	LOW WATER	FRESH WATER	RUNNING LIGHTS	NAMES OF CREW	GUESTS	
AT START TOOK ABOARD AT FINISH CONSUMED PRICE	PLACE TIME	PLACE TIME	TOOK ABOARD GALLONS	DISPLAYED EXTINGUISHED			

Fig. 75. Suggestions for motor boat log book sheet

CHAPTER XVIII

Steering

Propeller Working Ahead (Right Hand Propeller)

THE water which is drawn into the propeller from forward of it in a line parallel to the keel has no appreciable effect upon steering. However, water is thrown out from the after side of the propeller, more or less radially from the blades. This rotary current set up strikes against the rudder to a greater or less extent (depending upon the position of the rudder and amount of helm given it) and tends to throw the stern sideways. The upper blades, which are moving from port to starboard, throw their water against the upper portion of the rudder, and the lower blades drive their current against the lower starboard side of the rudder. If the lower part of the rudder is greater in area (relative to the center of motion of the water thrown radially away from the after side of the propeller) the resultant effect will be to throw the stern to port. If the upper area is greater, then the stern will be thrown to starboard. Thus it will be seen that the position and size of the rudder relative to that of the propeller will have a certain influence upon steering.

Propeller Working Astern

Water is drawn in from astern and forced out forward. The water forced out is thrown against the boat's underbody—that from the upper blades against the starboard side, and that from the lower against the port side. As the upper blades are working more effectively in this case, it follows that in backing, a boat's stern will be thrown to port.

Sidewise Force of Propellers

There is a certain sidewise force exerted by a propeller. The upper blades moving from port to starboard tend to force the stern to port, and the lower blades working from starboard to port have a tendency to throw the stern to starboard. As the lower blades are working in water of greater density, their action will have the greatest effect, with the result that the boat's stern will be thrown to starboard. In backing, the stern will be thrown to port.

This resultant action of the sidewise force of the propeller is quite pronounced in motor craft, especially when the upper blades are near the surface. Most motor boats have a tendency to work off their course to port for this reason. The effect is greatest when the boat is starting from rest, as then the tendency to "churn" the water is maximum when the boat's speed is minimum. For this reason it is essential and desirable to turn to port when starting up if it is desired to change the boat's course quickly. Many motor boats will not turn to starboard until they have considerable way on, while they will turn very readily to port.

Wake Current

The wake current, or that drawn along by the boat, is greatest at the stern of the boat at or near the surface of the water. It rapidly diminishes below the surface. The wake current is maximum at maximum speed of the boat, being zero when the boat is at rest. The influcnce of wake current, which has the greatest effect upon the upper blades, is to neutralize the greater sidewise effect of the lower blades when the boat speed is maximum.

Effect of the Propeller Upon Steering

1. Boat and Propeller Going Ahead

Here the rudder is the controlling factor for reasons explained, although the average motor boat tends to turn better to port. If the helm is put hard over when the boat is going full speed ahead, the first effect will be to throw the whole boat to the side opposite to which it is desired to go, the stern going off the most and not returning to the line of the original path of the boat until the bow has turned several points. The boat turns with increasing rapidity until she reaches a point from which she turns on a path which is practically a circle. As the boat swings around this circle, her bow is pointed inward, and her stern outward. The exact point on her keel which moves around on a true circle depends largely on the boat's speed—the faster the boat is, the nearer the bow will this point be.

The speed at which a boat is moving at the time her helm is put over has little effect upon her turning space, although the time of turning will be less with the faster boats. If the helm is put over to clear a stationary object, the speed will not be a factor in determining whether the object will be cleared or hit. However, speed will be a factor in the force with which the object is hit.

Generally speaking, a boat may be turned through eight points with a fore and aft "advance" of four boat lengths.

It is generally considered more safe to avoid a stationary object close ahead by means of reversing than by attempting to clear it by putting the helm hard over. The safest method is to put the helm hard to port, and, as soon as the bow bgeins to swing, to reverse, immediately putting the helm hard to starboard.

2. Boat and Propeller Going Astern

A. Boat Just Beginning to Back:

I. Helm amidship. Result: Stern moves to port.

II. Helm a-starboard. Result: Stern moves strongly to port.

III. Helm a-port. Result: Stern moves slowly to port.

B. Boat Gathers Speed Astern:

I. Helm amidship. Result: Stern moves to port.

II. Helm a-starboard. Result: Stern moves rapidly to port.

III. Helm a-port. Result: Stern moves slightly to starboard.

3. Boat Going Ahead, Propeller Astern

I. Helm amidships. Result: Bow swings to starboard.

II. Helm hard to starboard. Result: Bow swings to starboard.

III. Helm hard to port. Result: Uncertain.

If the boat has begun to swing from a hard over helm before the propeller is reversed, she will generally continue to swing when the propeller is reversed. If the propeller is reversed before the helm is put over, the above results will follow. One may be sure of results by first putting the helm hard over, then reversing the propeller, then reversing the helm; for example: When going ahead, if it is desired to throw the bow to

Port	*Starboard*
Stop propeller	Stop propeller
Starboard helm	Port helm
Reverse propeller	Reverse propeller
Port helm	Starboard helm

4. Boat Going Astern, Propeller Ahead

I. Helm amidships. Result: Uncertain.

II. Helm to port. Result: Stern swings decidedly to port.

III. Helm to starboard. Rseult: Stern will probably swing to starboard.

Note: To steer a straight course when backing have helm to port.

Table Showing Action of Boat Under Various Steering Conditions

Direction of motion of Boat	Ahead	Astern	Ahead	Astern
Direction of Motion of Propeller	Ahead	Astern	Astern	Ahead
Helm amidships	Boat will generally work to port slightly	Stern moves to port.	Bow swings to starboard	Stern will probably swing to port
Helm a-port	Boat swings to starboard	Stern moves slowly to port*	Uncertain	Stern swings strongly to port
Helm a-starboard	Boat swings to port.	Stern moves strongly to port	Bow swings to starboard	Stern may swing slightly to starboard

*As stern speed increases this motion may be changed to a slight starboard motion.

RECAPITULATION

From the foregoing it will be seen that it is practically impossible to make a boat's bow swing to port when backing. Therefore, in maneuvering from a position at rest it is best to plan to turn to starboard as follows:

1. Helm hard a-port.
2. Go ahead with propeller.
3. Go ahead as far as safe, swinging sharply to starboard, gaining as much headway as possible.
4. Reverse propeller full speed.
5. Immediately shift helm to hard a-starboard.
6. Back as far as possible at speed.
7. Propeller full speed ahead.
8. Put helm hard a-port at once.

If it is necessary to turn to port, proceed as follows:

1. Helm hard a-starboard.
2. Go ahead full speed until boat gathers good speed.
3. Stop propeller and let boat run.
4. Reverse propeller full speed.
5. Port helm immediately.
6. Run astern as far as possible.
7. Go ahead full speed and—
8. Put helm a-starboard immediately.

On boats of light draft, the whole conditions described above may be reversed, especially when the propeller is poorly or incompletely submerged.

CHAPTER XIX

Boat Equipment, Provisions and Supplies

(Navigation instruments not included)

Boat Gear

Extra heavy anchor
Small anchor
Sea anchor
Anchor chain
Emergency sail, mast and tackle
Emergency hand tiller
Supply of different sizes of small line, marline and other cordage
200 feet one-inch (diameter) line
Bow and stern lines
Line for halyards
Supply of canvas or tarpaulin
Bow, stern and running lights
Riding light
Lanterns
Fog bell
Whistle
Fog-horn
Bow and stern staffs
Life preservers
Ring buoys
Fenders
Fire extinguisher
Bilge pump
Extra oars and oarlocks for dinghy
Broom, mop and bucket
Boat hook
Cushions

Provisions

Matches
Flour
Sugar
Pancake flour (self-raising)
Baking powder
Coffee
Tea
Cocoa
Dried peas, beans and prunes
Rice
Macaroni
Cereals and breakfast foods
Cornstarch, tapioca and jello
Beef cubes
Bread
Cheese
Salt
Pepper
Soda
Cinnamon
Sage
Mustard
Curry
Paprika
Fresh milk in jars
Ham and bacon
Salt pork
Butter
Eggs
Lard
Catsup
Worcestershire sauce
Horse radish
Pickles and olives
Olive oil
Vinegar
Preserved fruits and jellies
Several boxes of soda, graham and oyster crackers
Bottle of syrup
Canned soups, beans, corn, tomatoes and peas

Provisions (*Continued*)

Canned roast beef, corned beef, veal loaf and tongue, potted ham
Canned sardines, tuna fish, salmon, crab-flakes and lobster
Canned milk and cream
Jars of sliced bacon, smoked beef and codfish
Potatoes, onions, carrots and such other fresh vegetables
Oranges, lemons, bananas and other fruits
Fresh water (in stone jugs or tanks)
Liquors
Soft beverages
Alcohol
Kerosene

Miscellaneous Supplies

Blankets
Oil skins
Bathing suits
Fishing tackle
Fire arms
Water-proof bag
Camera and photographic supplies
Smoking material
Sewing and mending kit
Medicine kit
Tooth-brushes and powder
Graphaphone and records
Playing cards
Writing paper
Cook book
Pocket flash-light
Ball of twine
Candles
Soap
Scouring and soap powders
Ammonia
Brass polish
Dish cloths, towels and mop
Face and bath towels
Wash-cloths
Table-cloths or oil-cloth
Napkins
Table
Chairs
Table silver
Dust pan
Clothes brush
Hatchet
Supply of assorted nails and screws
Hydrometer for storage battery
Supply of extra electric light bulbs
Caulking iron
Marline spike
Palm

China and Glassware

Eight dinner plates
Eight lunch plates
Eight soup plates
Eight bread and butter plates
Eight cups and saucers
Two platters
Four open bowls, various sizes
Two covered dishes
One cream pitcher
One sugar bowl
Several small flat dishes
Large pitcher
Twelve water glasses
Twelve high-ball glasses
Twelve wine glasses
Three decanters
One thermos carafe
Set of salt and pepper shakers
Syrup jar

Galley Utensils

- Stove and fuel
- Fireless cooker
- Broiler or toaster
- Can opener
- Coffee pot
- Tea pot
- Coffee strainer
- Two cooking spoons
- Cooking fork
- Cooking knife
- Potato knife
- Bread knife
- Carving set
- Ice pick
- Small dish pan
- Small hand basin
- Pancake turner
- Egg beater
- Lemon squeezer
- Good sized tea-kettle
- Two galvanized water pails
- One large and one small preserving kettle
- One large and one small stewing or sauce pan
- Two large and one small skillet or frying pan
- One double boiler
- Pancake griddle
- Three baking or bread pans
- Six pie plates
- Bottle opener
- Cork-screw
- Sieve
- Potato masher
- Measuring cup
- Meat chopper
- Cocktail shaker

Tools and Engine Accesories

- Ten-inch mill file
- Eight-inch half round file
- Rat-tail file
- File handle
- Cotter pin extractor
- Cold chisel, ⅝-inch
- Cape chisel
- Round-nose chisel
- Small hand vise
- Pair gas pliers
- Half-dozen spark plugs, each in separate container—generally a round, screw top box
- Vibrator complete for coil in separate container
- Spare union, R. & L., for gas pipe line
- Large solid metal screwdriver
- Small screwdriver
- Six-inch combination pliers (with cutting edge)
- S wrenches ¼ to ⅝ opening (three in all)
- Small monkey wrench
- Small Stilson wrench
- Socket wrench to fit spark plugs
- Eight-oz. machinist's hammer
- Eight-inch Stilson wrench
- Eight-inch monkey wrench
- Six-inch Stilson pipe wrench
- Fourteen-inch Stilson pipe wrench
- Pair 10-inch gas pliers
- One-inch patent combination pliers
- Ten-inch flat file
- Seven-inch three-corner-file
- Two screwdrivers, one 6-inch and one 10-inch or 12-inch
- Nail hammer, and few assorted nails
- Six-inch monkey wrench

Tools and Engine Accessories (*Continued*)

- Cold chisel, ½-inch wide and about 6 inches long
- Twelve-inch monkey wrench
- Spool of fine copper wire
- Three to four feet of spring brass wire, about No. 12 gauge
- Piece of emery cloth
- Piece wire-inserted packing, big enough for cylinder head
- Ball candle wicking
- Small can Smooth-on for cracked water jackets
- Small tin shellac, for gas pipe line
- One can each of assorted cotter pins, assorted hex nuts, assorted washers. (These are standard goods and may be had at any supply house)
- One valve complete (in separate container)
- Two spare valve springs
- Emery paste for grinding in valves
- Hack saw and twelve blades
- Solder, copper and bundle of flux solder (solder with resin core)
- No. 2 roll friction tape
- Revolution counter
- Oil gun
- White lead
- Graphite
- Volt ammeter
- Spare pipe fittings
- Extra batteries
- Extra parts
- Gasoline
- Hard grease
- Lubricating oil
- Rubber hose
- Spark plugs
- Tool kit
- Waste
- Asbestos packing
- Insulated wire
- Pint measure
- Funnel
- Tire tape

Navy Department's Suggestion for Medical Kit

Antiseptic tablets (bichloride mercury)	1 bottle
Bandages	6
Beef, extract	1 bottle
Calomel, tablets	1 bottle
Cathartic tablets	1 bottle
Chlorodyne tablets	1 bottle
Gauze	2 yards
Lead and opium tablets	1 bottle
Mustard plasters	1 box
Packages, first-aid	6
Plaster, rubber	1 roll
Quinine pills (3-grain)	1 bottle
Soda, bicarbonate	1 can
Tourniquets, rubber, instant	4
Vaseline, carbolized	1 jar
Whisky	1 bottle

Directions for giving medicines.

CHAPTER XX

Suggestions for Meals

Breakfasts

1

Stewed fruit
Cereal with butter and sugar
Corn fritters Coffee

2

Bacon and eggs
Fried potatoes
Coffee

3

Cereal fried with bacon
Lazy biscuits Coffee
Eggs

4

Scrambled eggs
Canned corn and bacon
Corn bread Coffee

5

Hominy grits browned with bacon
Stewed Fruit Coffee

Dinners

1

Beef pilau
Crackers
Blackberry dumplings
Coffee

2

Salmon chowder
Hot pilot bread
Stewed loganberries
Coffee

3

Bean puree
Spinach with eggs
Coffee Crackers
Candy

A COMPANY DINNER

Mock turtle soup
Cold fish with mayonnaise Lazy biscuits
Okra sliced with tomatoes
Deep blueberry pie
Coffee

Suppers

1

Baked beans with tomato sauce
Crackers Tea
Apricots

2

Corn soup
Biscuits Tea
Fresh fruit

3

Lamb stew with peas

Crackers Tea

4

Hulled corn and fresh milk

Ginger cookies

Tea

5

Mock turtle and tomato soup

Pilot bread

Tea

6

Fish hash of salt cod and smoked salmon with potatoes

Prunes Tea

Recipes

Following are the recipes of a few dishes which the most masculine amateur cook can successfully prepare:

Lamb and Peas

One can lamb.
One can peas.
Two boiled potatoes.
One onion.
Stock.

Disc lamb and potatoes and cook all together with enough stock or water to moisten thoroughly. Season and serve.

Beef Pilau

One can beef.
One can tomatoes.
One tablespoon butter.
One onion.
One-half cup rice.

Brown onion lightly in butter; add the beef, sliced; season with salt and pepper. Cook the washed rice for five minutes, then add to meat with the tomato and one cup of water. Simmer until rice is tender.

Fricasseed Beef

One can beef.
One can soup stock or a bouillon cube in water.

Brown an onion minced in butter, add the soup stock, season highly and thicken with flour and butter. Cook the meat slices in this until hot and well seasoned.

CHAPTER XXI

Navy Signaling

(From Navy Deck and Boat Book)

THE flags and pennants used in transmitting the United States Navy Flag Code are as follows:

Alphabet Fags—Negative flag (alphabet flag K). Preparatory flag (alphabet flag L). Annulling flag (alphabet flag N). Interrogatory flag (alphabet flag O). Affirmative flag (alphabet flag P). The following special flags are used: Numeral flag, Repeaters, Danger and designating flag, Answering and divisional point pennant, Cornet, Call flags, Ship call pennants, Indicators.

The alphabet flags are the same as those of th International Code. The letters E and T are not used singly because of their display by the Ardois system as night speed indicators, and because with the whistle they indicate a change of course. The letter I is used singly as the dispatch and breakdown flag.

To prevent confusion and mistakes due to the similarity in the sounds of different letters when calling out flags or recording signals, the words given in the following table are used as the names of the letters on board of all vessels and at all times:

A	Able	N	Nan
B	Boy	O	Oboe
C	Cast	P	Pup
D	Dog	Q	Quack
E	Easy	R	Rush
F	Fox	S	Sail
G	George	T	Tare
H	Have	U	Unit
I	Item	V	Vice
J	Jig	W	Watch
K	Kink	X	X Ray
L	Love	Y	Yoke
M	Mike	Z	Zed

Negative Flag K

The negative flag (alphabet flag K) when hoisted in answer to a signal means "Not granted" or "No." The call of the boat to which it is addressed as an answer shall be displayed over it in order to avoid any chance of a misunderstanding. Hoisted over a signal it puts the message in a negative sense.

Preparatory Flag L

The preparatory flag (alphabet flag L) hoisted over a signal means "Prepare to execute the signal now shown as soon as the signal of execution is made.

The signal of execution is the starting from its point of hoist the same signal (without the preparatory flag) or some other signal relating to the same movement or maneuver. Thus, if the signals were made to prepare to moor ship, the signal of execution might be the hauling down later of a signal to "anchor in succession in inverted order."

Annulling Flag N

The annulling flag (alphabet flag N) annuls all signals at that moment displayed on the same mast. In this case only it is to be answered by hauling down all answering pennants which may have been hoisted in reply to the signal or signals. In case ships have the signal or signals hoisted they shall also display the annulling flag and haul all down with their hauling down on the flagship. Hoisted alone, it annuls the last signal made or the last hoist.

Any signal previously made may be annulled by hoisting the signal again with the annulling flag either over it or hoisted at the same time.

Interrogatory Flag O

The interrogatory flag (alphabet flag O) when hoisted over a signal changes its meaning to the interrogatory form.

The single display of the interrogatory in answer to a signal means that the signal cannot be read or is not understood.

The interrogatory hoisted alone by a flagship means "You are repeating the signal wrong," or "What movement are you making?" according to the circumstances which will be evident.

Affirmative Flag P

The affirmative flag (alphabet P) when hoisted in answer to a signal, means assent, consent, permission granted, or "Yes." The call of the boat to which it is addressed as an answer shall be displayed over it in order to avoid any chance of a misunderstanding.

Hoisted over a signal, it means that the specific work or service called for by that signal has been completed or the duty has been performed. For example, the signal "Moor," with the affirmative over it, means "I have moored."

It is hoisted alone when getting under way with other vessels in formation; when ready to steam ahead after "Man overboard" or other contingency involving stopping; in mooring ship; and in other cases to indicate that some duty called for in a previous signal has been completed.

Numeral Flag

The numeral flag hoisted over certain alphabet flags indicates that those flags are to be interpreted as numerals. The flags whose meanings are thus changed are as follows:

Q	1	V	6
R	2	W	7
S	3	X	8
T	4	Y	9
U	5	Z	0

Repeaters

The repeaters serve to reproduce, in numeral and vocabulary signals only, the alphabet flags hoisted above them. The first repeater reproduces the first alphabet flag, the second repeater the second, and the third the third.

Danger and Designating Flag

The danger and designating and Navy-Register-Use flag hoisted alone indicates danger ahead. A compass signal under it indicates the bearing of the danger from the boat making the signal.

Used as a designating flag it designates a particular boat, place, person, or thing, when hoisted over or at the same time and if possible, on the same mast as the signal representing the object referred to. For a boat, the signal will be her call or her International Signal letters; for a place or thing, the signal in the vocabulary; for an officer on the active list, his signal number in the Navy Register; for an officer on the retired list, or an enlisted man or other person, the signals spelling out his name. In making an officer's number the January Navy Register of each year shall be used on and after July 1 of that year and until and including June 30 of the following year.

Answering Pennant

The answering pennant is hoisted where it can be best seen—at the truck, gaff, or yardarm—in answering, and kept there until the signal is hauled down. At sea, if displayed at the yardarm it shall be at the side not occupied by the speed cone unless that leads to concealment by smoke. In port it shall be displayed at the starboard yardarm.

In order that there shall be no uncertainty as to the signal that is answered by a boat, the latter shall display the answering pennant under the call of the boat making the signal; except that in answering a signal from the senior flagship the latter's call shall not be displayed.

The answering pennant is used as a divisional point to represent the divisions of mixed quantities referred to by a signal made at the same time or just previously.

Cornet

The cornet hoisted at the fore, or at the highest gaff of signal yard if the foremast head cannot be used for its display, is a peremptory order for all officers and others absent from the boat to repair on board at once. A gun fired denotes urgency.

The cornet hoisted half yardarm high is a call for the whole force to receive a semaphore or wigwag message.

Union Jack

The Union Jack hoisted at the fore is a signal for a pilot. Hoisted at a yardarm it indicates that a general court-martial or court of inquiry is in session on board. In port a gun shall be fired when it is hoisted upon the meeting of the court.

Powder Flag

The powder flag (alphabet flag B) shall be displayed at the fore on all vessels while taking on board or discharging explosives or loaded projectiles, or fuel oil or gasoline in large quantities, and in the bows of all boats and lighters transporting the same. It is also to be displayed by a ship engaged in target practice with either guns or torpedoes, while the firing is in actual progress. It shall be hauled down halfway when off the firing line if the practice is to continue, and hauled down at "Cease firing" or "Secure." It may also be used when standardizing propellers, to indicate when the boat is on the course and observations are in progress.

Dispatch and Breakdown Flag

The dispatch and breakdown flag (alphabet flag I) shall be worn at the main in all dispatch vessels to indicate the nature of their service. No vessel shall hoist the dispatch flag without proper authority, or display it as such until actually under way and out of formation; but when it is hoisted she shall not be interfered with by an officer junior to the one by whom she is sent on such service, except when the public interests imperatively demand such action, of the necessity for which the senior officer present must be the responsible judge. A vessel engaged in carrying dispatches or orders through a fleet should hoist below the dispatch flag the call of the boat to which she is next bound; or she may display in inverse order the calls of all the boats to be communicated with, that of the boat to be next communicated with being the last one in the hoist, and as soon as any boat has been communicated with her call shall be detached from the hoist.

In formation under way this flag shall be kept rounded up "in stops" at the fore ready to break in case of accident to machinery or steering gear. When a guide flag is displayed at the foremast head the breakdown flag shall be hoisted "in stops" below the guide flag ready to break below it. In case of accident which is likely to necessitate slowing down or leaving the formation, it shall be instantly broken as an emergency signal, and implies "breakdown" or "not under control," and other boats must keep clear until the boat displaying the breakdown flag shall have gotten well clear of the formation. The breakdown flag does not relieve a boat from responsibility in cases of collision, even though she may have sheered out of formation. It shall be kept flying during daylight until repairs are completed and the ship is headed for the formation to resume her position, or until she is beyond signal distance.

In case of man overboard, a ship in formation shall break this flag and at once lower it part way (but not below the level of the smokestacks) as a signal for "man overboard."

Guard and Guide Flag

The guard and guide flag when used at anchor is a guard flag, and indicates that the boat upon which it is hoisted is charged with the guard duty of the division, squadron, or force then at anchor within signal distance. It shall be hoisted at the fore between sunrise and sunset. From sunset until sunrise a red light shall be displayed at the fore truck, except under such conditions as may cause it to be confused with navigation lights. The guard flag shall not be hoisted nor the red light shown, by a flagship or vessel of the senior officer present to indicate that she has the guard duty. A boat may be directed to take the guard duty by hoisting the guard flag under the boat's call; this is answered by hoisting the guard flag. A boat is relieved from guard duty by hoisting the guard flag over the boat's call; this is answered by hauling down the guard flag. To call a guard boat alongside, the flagship may hoist the guard flag under the boat's call or under the call of the division to which the boat belongs; or at night may display a red light after the call, as above. This is answered by dipping the guard flag or pulsating the red light. Boats on guard duty shall display a small guard flag from a staff in their bows during daylight.

When a ship is under way and in formation the display of this flag at the fore denotes that she is the guide in the formation. The guide flag at the dip (lowered part way) indicates that the guide boat is temporarily out of position. The division guide pennant (position pennant) shall be displayed under the same rules for rear divisions or squadrons. A boat may be directed to take the guide by hoisting the guide flag under the boat's call, or to cease being the guide by hoisting it over the boat's call. This is answered by hoisting or hauling down the guide flag or pennant as called for.

Miscellaneous Flags

The convoy and position and division guide pennant shall be worn at the fore of all ships on escort duty to indicate the nature of their service. No vessel shall hoist the convoy pennant without proper authority; but when it is hoisted she shall not be interfered with by an officer junior to the one by whom she was sent on such service, except when the public interests imperatively demand such action, of the necessity for which the senior officer present must be the responsible judge.

It is also a division (or squadron) guide pennant, and in compound formations is worn by the guides of divisions (or squadrons) other than the fleet guide.

A boat in formation not on escort duty hoists this pennant to indicate that she has attained an assigned position; but it shall not be used in tactical evolutions except as prescribed under fleet tactics. When a boat in formation has been in position and loses it, she shall hoist the position pennant at half-mast as an indication of the fact, and keep it so until she regains her position, when she shall run it all the way up and immediately haul it down. Under battle conditions it shall not be used for this purpose.

The position pennant displayed under a call by the senior officer, means to the boat signaled: "You are out of position," or "You are out of order."

The full-speed and meal and flag-officer-leaving pennant is kept hoisted as a single display at the port yardarm during the period when the crew is at meals on board vessels at anchor, whether or not the colors are hoisted. This pennant shall be shown for this purpose without reference to the flagship. Under way in formation it shall be used as a full-speed pennant on the same side as the speed cone. When a flag officer is about to leave his flagship officially during the day, this pennant shall be displayed directly under the flag and hauled down when he shoves off.

The general- and boat-recall flag is a peremptory order for all boats absent from that ship on detached duty, or otherwise, to return with all speed to their ship. A numeral displayed below the general recall is an order that all boats except the one indicated shall return to the ship. The general-recall flag under a numeral indicates the recall of the ship's boat having that numeral as a call number. Commanding officers of all ships shall assign numbers to all boats for this and for boat exercise purposes. A recall shall be kept hoisted until the boat is made out as returning in obedience to the signal. If necessary, but only in case of emergency, the ship's call letter may be sounded on the steam whistle or a gun fired to attract the attention of the boat. When a boat recall is hoisted with the annulling flag over it, it indicates that the boat so designated shall not return to the boat at the time previously prescribed, but shall wait for the further display of her recall.

At night a boat may be recalled by the display of I, followed by the number of the boat, and by the boat's call if there can be any uncertainty as to which ship she is signaling; or special night recalls may be assigned for particular occasions, observing due care that the recall used shall not be such as to render it possible to be mistaken by other ships as a signal to them.

The quarantine flag (alphabet flag Q) is hoisted at the foremast head or most conspicuous hoist on all infected ships or ships in quarantine. It shall be kept flying day and night and be carried in the bow of all boats belonging to the ship having this flag hoisted. It should be hoisted by incoming ships as a signal to the health officer of the port that pratique is desired.

The battle efficiency pennant shall be worn at the fore when at anchor on such vessels as may have been officially declared entitled to fly it for excellence in battle efficiency.

The Red Cross flag is, by international agreement, to be worn at the fore on all hospital vessels. It shall also be displayed over the field hospital of any naval force on shore and on hospital boats of landing parties.

The submarine warning flag is hoisted on the tender or parent ship of submarines or on launches accompanying them, to indicate that submarines are operating submerged in that vicinity.

The church pennant shall be hoisted at the same place of hoist and over the ensign during the performance of divine service on board vessels of the Navy.

The battalion flags for infantry and artillery are provided for naval landing forces.

The Naval Militia distinguishing flag shall be hoisted at the mast-head (fore when there is more than one mast) at all times on all vessels loaned by the Navy Department to a State for the use of the Naval Militia or Naval Reserve, and on all vessels "in commission in reserve" and assigned to the States for the instruction of the Naval Militia or Naval Reserve, when such vessels are actually under the command of a Naval Militia or Naval Reserve officer.

The Naval Militia commission pennant and commodore's pennant shall be worn by the Naval Militia vessels in the same way that is prescribed for the similar pennants on vessels of the Navy.

To Signal the Engineer

When engine is stopped, 1 bell for Ahead Slow.
When running ahead slow, jingle for Full Speed Ahead.
When running full speed ahead, 1 bell for Slow Down.
When running ahead slow, 1 bell for Stop.
When stopped, 2 bells for Astern Slow.
When running astern slow, jingle for Full Speed Astern.
When running astern slow or at full speed, 1 bell for Stop.
When running full speed ahead, 4 bells for Full Speed Astern.

When no jingle bell is provided, use the following signals:

When stopped, 1 bell for Ahead.
When running ahead, 1 bell for Stop.
When stopped, 2 bells for Astern.
When running astern, 1 bell for Stop.
When running ahead, 4 bells for Astern.

CHAPTER XXII

Determining Compass Deviation

IN Chapter VIII (Pages 61, 62, 65, 66 and 67), several methods were referred to which can be used in determining the deviation of one's compass on his boat. On page 62 was mentioned the method of putting a boat over a number of courses whose magnetic direction could be determined from a chart.

Fig. 76 shows a number of such courses laid out from Government chart (No. 369, Long Island Sound) of Huntington Bay. On this bay there are several aids to navigation shown on the chart such as bell buoy, No. 13A, red and black spar off Eaton's Neck, can buoy No. 13, Eaton's Point Lighthouse, etc., etc. All of these aids are near enough to each

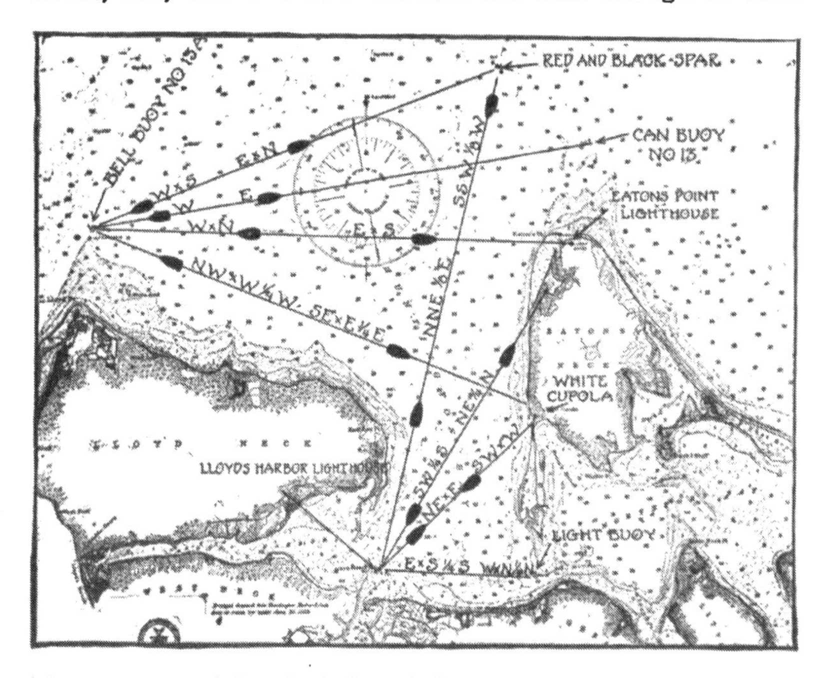

Fig. 76. Determining deviation of the compass by putting the boat over a number of courses, between aids to navigation, whose magnetic direction can be determined from the chart

LONGITUDE

DATE	92° 77°		91° 76°		90° 75°		89° 74°		88° 73°		87° 72°		86° 71°		85° 70°		84° 69°		83° 68°		82° 67°		81° 66°		80° 65°		79° 64°		78° 63°		77° 62°		76° 61°
	Sub.	Sub.	Sub.	Add	Add	Add	Add	Add	Add	Add	Add	Add	Add	Add	Add	Add	Add	Add	Add	Add	Add	Add	Add	Add	Add	Add	Add	Add	Add	Add	Add	Add	Add
MAY 1	5	3	1	1	3	5	7	9	11	13	15	17	19	21	23	25	27	29	31	33	35	37	39	41	43	45	47	49	51	53	55	57	59
7	4	2	0	2	4	6	8	10	12	14	16	18	20	22	24	26	28	30	32	34	36	38	40	42	44	46	48	50	52	54	56	58	60
22	5	3	1	1	3	5	7	9	11	13	15	17	19	21	23	25	27	29	31	33	35	37	39	41	43	45	47	49	51	53	55	57	59
JUNE 1	6	4	2	0	2	4	6	8	10	12	14	16	18	20	22	24	26	28	30	32	34	36	38	40	42	44	46	48	50	52	54	56	58
7	7	5	3	Sub. 1	1	3	5	7	9	11	13	15	17	19	21	23	25	27	29	31	33	35	37	39	41	43	45	47	49	51	53	55	57
12	8	6	4	2	0	2	4	6	8	10	12	14	16	18	20	22	24	26	28	30	32	34	36	38	40	42	44	46	48	50	52	54	56
17	9	7	5	3	Sub. 1	Sub. 1	3	5	7	9	11	13	15	17	19	21	23	25	27	29	31	33	35	37	39	41	43	45	47	49	51	53	55
22	10	8	6	4	2	0	2	4	6	8	10	12	14	16	18	20	22	24	26	28	30	32	34	36	38	40	42	44	46	48	50	52	54
26	11	9	7	5	3	1	1	3	5	7	9	11	13	15	17	19	21	23	25	27	29	31	33	35	37	39	41	43	45	47	49	51	53
JULY 1	12	10	8	6	4	2	0	2	4	6	8	10	12	14	16	18	20	22	24	26	28	30	32	34	36	38	40	42	44	46	48	50	52
7	13	11	9	7	5	3	Sub. 1	1	3	5	7	9	11	13	15	17	19	21	23	25	27	29	31	33	35	37	39	41	43	45	47	49	51
14	14	12	10	8	6	4	2	0	2	4	6	8	10	12	14	16	18	20	22	24	26	28	30	32	34	36	38	40	42	44	46	48	50
AUG. 5	13	11	9	7	5	3	1	1	3	5	7	9	11	13	15	17	19	21	23	25	27	29	31	33	35	37	39	41	43	45	47	49	51
15	12	10	8	6	4	2	0	2	4	6	8	10	12	14	16	18	20	22	24	26	28	30	32	34	36	38	40	42	44	46	48	50	52
20	11	9	7	5	3	1	Add 1	3	5	7	9	11	13	15	17	19	21	23	25	27	29	31	33	35	37	39	41	43	45	47	49	51	53
24	10	8	6	4	2	0	2	4	6	8	10	12	14	16	18	20	22	24	26	28	30	32	34	36	38	40	42	44	46	48	50	52	54
27	9	7	5	3	1	Add 1	3	5	7	9	11	13	15	17	19	21	23	25	27	29	31	33	35	37	39	41	43	45	47	49	51	53	55
30	8	6	4	2	0	2	4	6	8	10	12	14	16	18	20	22	24	26	28	30	32	34	36	38	40	42	44	46	48	50	52	54	56
SEPT. 3	7	5	3	1	Add 1	3	5	7	9	11	13	15	17	19	21	23	25	27	29	31	33	35	37	39	41	43	45	47	49	51	53	55	57
6	6	4	2	0	2	4	6	8	10	12	14	16	18	20	22	24	26	28	30	32	34	36	38	40	42	44	46	48	50	52	54	56	58
9	5	3	1	Add 1	3	5	7	9	11	13	15	17	19	21	23	25	27	29	31	33	35	37	39	41	43	45	47	49	51	53	55	57	59
12	4	2	0	2	4	6	8	10	12	14	16	18	20	22	24	26	28	30	32	34	36	38	40	42	44	46	48	50	52	54	56	58	60
15	3	1	Add 1	3	5	7	9	11	13	15	17	19	21	23	25	27	29	31	33	35	37	39	41	43	45	47	49	51	53	55	57	59	61
18	2	0	2	4	6	8	10	12	14	16	18	20	22	24	26	28	30	32	34	36	38	40	42	44	46	48	50	52	54	56	58	60	62
21	1	Add 1	3	5	7	9	11	13	15	17	19	21	23	25	27	29	31	33	35	37	39	41	43	45	47	49	51	53	55	57	59	61	63
23	0	2	4	6	8	10	12	14	16	18	20	22	24	26	28	30	32	34	36	38	40	42	44	46	48	50	52	54	56	58	60	62	64
26	Add 1	3	5	7	9	11	13	15	17	19	21	23	25	27	29	31	33	35	37	39	41	43	45	47	49	51	53	55	57	59	61	63	65
29	2	4	6	8	10	12	14	16	18	20	22	24	26	28	30	32	34	36	38	40	42	44	46	48	50	52	54	56	58	60	62	64	66
OCT. 2	3	5	7	9	11	13	15	17	19	21	23	25	27	29	31	33	35	37	39	41	43	45	47	49	51	53	55	57	59	61	63	65	67
6	4	6	8	10	12	14	16	18	20	22	24	26	28	30	32	34	36	38	40	42	44	46	48	50	52	54	56	58	60	62	64	66	68
9	5	7	9	11	13	15	17	19	21	23	25	27	29	31	33	35	37	39	41	43	45	47	49	51	53	55	57	59	61	63	65	67	69
13	6	8	10	12	14	16	18	20	22	24	26	28	30	32	34	36	38	40	42	44	46	48	50	52	54	56	58	60	62	64	66	68	70
17	7	9	11	13	15	17	19	21	23	25	27	29	31	33	35	37	39	41	43	45	47	49	51	53	55	57	59	61	63	65	67	69	71
23	8	10	12	14	16	18	20	22	24	26	28	30	32	34	36	38	40	42	44	46	48	50	52	54	56	58	60	62	64	66	68	70	72
NOV. 14	7	9	11	13	15	17	19	21	23	25	27	29	31	33	35	37	39	41	43	45	47	49	51	53	55	57	59	61	63	65	67	69	71
20	6	8	10	12	14	16	18	20	22	24	26	28	30	32	34	36	38	40	42	44	46	48	50	52	54	56	58	60	62	64	66	68	70
24	5	7	9	11	13	15	17	19	21	23	25	27	29	31	33	35	37	39	41	43	45	47	49	51	53	55	57	59	61	63	65	67	69
27	4	6	8	10	12	14	16	18	20	22	24	26	28	30	32	34	36	38	40	42	44	46	48	50	52	54	56	58	60	62	64	66	68
30	3	5	7	9	11	13	15	17	19	21	23	25	27	29	31	33	35	37	39	41	43	45	47	49	51	53	55	57	59	61	63	65	67
DEC. [illegible]	2	4	6	8	10	12	14	16	18	20	22	24	26	28	30	32	34	36	38	40	42	44	46	48	50	52	54	56	58	60	62	64	66
[illegible]	1	3	5	7	9	11	13	15	17	19	21	23	25	27	29	31	33	35	37	39	41	43	45	47	49	51	53	55	57	59	61	63	65
[illegible]	0	2	4	6	8	10	12	14	16	18	20	22	24	26	28	30	32	34	36	38	40	42	44	46	48	50	52	54	56	58	60	62	64

(Courtesy of A. McNevin)

Fig. 77. The amounts which must be added to or subtracted from standard (clock) time to obtain real (sun) time

other to permit their being seen from a boat in any position on the bay. Consequently, it is possible to lay off eight courses between them which with the reverse courses will give the motor boatman sixteen courses whose magnetic direction may be definitely determined from the chart as shown in Fig. 76. By putting the boat over each of these sixteen courses, and observing the direction as shown by compass on the boat, and then comparing these compass headings with the actual magnetic courses shown by the chart (Fig. 76) one can readily determine his compass deviation on the various headings.

The Sun Pelorus

On page 65 (and Fig. 40) is described a method of determining deviation by means of the sun compass or shadow pelorus. Figures 77 and 78 will give the necessary data for setting the sun pelorus for determining deviation. The equation of time for the various longitudes and days of the year is given in Fig. 77, and represents the number of seconds which must be added to or subtracted from the standard (clock) time to obtain the "real" (sun) time. The "real" time must be used with Fig. 78 for determining the setting of the sun compass.

Bearings on a Near Object

Fig. 79 represents a very convenient method of determining deviation by means of bearings on an object on shore. Fig. 41 illustrated a method of using a distant object which could be located on the chart, but Fig. 79 represents an object (*A*) close at hand on shore, such as a tree, stake or a mark specially placed.

The boat is anchored in a fixed position, preferably with several anchors out bow and stern, so that she will swing only as desired. The compass is then taken ashore, and a bearing taken from *A* to the boat. As there will be no deviation to the compass on shore, a correct magnetic bearing can be obtained from *A* to the anchored boat. Then with the compass in its correct position on the boat, a bearing is taken with it from the boat to the mark *A* on shore. The difference in the two bearings will be the deviation of the compass on that particular heading which the boat is then on. (That heading shown by the compass lubberline corrected for the amount of deviation just determined.)

For example, if the bearing from the mark *A* on shore to the boat (as indicated with the compass on shore) is W N W, then the correct magnetic bearing from the boat to *A* will be

TRUE BEARINGS OF THE SUN
LATITUDE 40°

AM	March 16	April 3	April 16	May 1	May 16	June 15	Aug 2	Aug 16	Aug 30	Sept 18	Sept 23	Oct 1	Oct 14	Nov 3	Nov 14	Dec 16	Feby 2	Feby 15	March 3	PM
VIII.	112	106	102	97	94	90	95	99	103	109	110	113	116	121	123	127	123	120	116	IV.
10	114	108	104	99	96	91	97	100	105	111	112	115	118	123	125	128	124	122	118	50
20	116	110	106	101	97	93	98	102	107	113	114	117	120	125	127	130	126	124	119	40
30	118	112	108	103	99	95	100	104	109	115	116	119	122	127	129	132	128	126	121	30
40	120	114	110	105	101	97	102	106	111	117	118	121	124	129	131	134	130	128	124	20
50	122	116	112	107	103	98	104	108	113	119	120	123	126	131	133	136	132	130	126	10
IX	124	119	114	109	105	100	106	110	115	121	123	125	128	133	135	138	134	132	128	III
10	127	121	117	112	107	102	108	113	118	123	125	127	131	135	137	140	136	134	130	50
20	129	123	119	114	110	105	111	115	120	126	127	130	133	137	139	142	138	136	132	40
30	131	126	122	117	112	107	113	118	123	128	130	132	135	140	141	144	141	138	135	30
40	134	129	124	119	115	109	116	120	125	131	133	135	138	142	144	146	143	141	137	20
50	137	131	127	122	117	112	119	123	128	134	135	137	140	144	146	148	145	143	140	10
X	139	134	130	125	120	115	122	126	131	137	138	140	143	147	148	150	147	146	142	II
10	142	137	133	128	124	118	125	129	134	140	141	143	146	149	151	153	150	148	145	50
20	145	141	137	132	127	121	128	133	138	143	144	146	149	152	153	155	153	151	148	40
30	148	144	140	135	131	125	132	136	141	146	147	149	151	154	156	157	155	154	151	30
40	152	147	144	139	135	129	136	140	145	149	150	152	154	157	158	160	158	156	154	20
50	155	151	148	144	139	134	140	145	149	152	154	155	157	160	161	162	160	159	157	10
XI.	158	155	152	148	144	139	145	149	153	156	157	159	160	163	163	164	163	162	160	I
10	162	159	156	153	149	144	150	154	157	160	161	162	164	165	166	167	166	165	163	50
20	165	163	161	158	155	151	156	159	161	164	165	166	167	168	169	170	169	168	167	40
30	169	167	165	163	161	157	161	164	166	168	168	169	170	171	172	172	171	171	170	30
40	173	171	170	169	167	163	167	169	170	172	172	173	173	174	174	175	174	174	173	20
50	176	176	175	174	173	172	174	174	175	176	176	176	177	177	177	177	177	177	177	10

In North Latitude the Azimuth is reckoned from North: to East when time is A.M., to West when time is P.M.
In South Latitude the Azimuth is reckoned from South: to East when time is A.M., to West when time is P.M.

TRUE BEARINGS OF THE SUN
LATITUDE 44°

AM	March 16	April 3	April 16	May 1	May 16	June 15	Aug 2	Aug 16	Aug 30	Sept 18	Sept 23	Oct 1	Oct 14	Nov 3	Nov 14	Dec. 16	Feby 2	Feby 15	March 3	PM
VIII	113	108	104	100	97	93	97	101	105	110	112	114	117	122	124	127	123	121	116	IV.
10	115	110	106	102	98	95	99	103	107	112	114	116	119	124	126	129	125	123	119	50
20	117	112	108	104	100	96	101	105	109	114	116	118	121	126	128	131	127	125	121	40
30	119	114	110	106	102	98	103	107	111	117	118	120	123	128	130	132	129	127	123	30
40	122	117	113	108	105	100	106	109	113	119	120	122	126	130	132	134	131	129	125	20
50	124	119	115	111	107	103	108	111	116	121	122	124	128	132	134	136	133	131	127	10
IX.	126	121	117	113	109	105	110	114	118	123	125	127	130	134	136	138	136	133	129	III.
10	128	124	120	115	111	107	112	116	121	126	127	129	132	136	138	140	137	135	132	50
20	131	126	122	118	114	109	115	119	123	128	130	132	135	138	140	142	139	137	134	40
30	133	129	125	120	117	112	118	121	126	131	132	134	137	141	142	145	142	140	136	30
40	136	131	128	123	119	115	120	124	128	133	135	137	139	143	144	147	144	142	139	20
50	139	134	131	126	122	118	123	127	131	136	137	139	142	145	147	149	146	144	141	10
X.	141	137	134	129	125	121	126	130	134	139	140	142	145	148	149	151	149	147	144	II.
10	144	140	137	133	129	124	130	133	137	142	143	145	147	150	151	153	151	149	147	50
20	147	143	140	136	132	127	133	137	141	145	146	148	150	153	154	156	154	152	150	40
30	150	147	143	140	136	131	137	140	144	148	149	151	153	155	156	158	156	155	152	30
40	153	150	147	143	140	135	141	144	148	151	152	153	156	158	159	160	159	157	155	20
50	156	153	151	147	144	139	145	148	151	154	156	157	158	161	161	163	161	160	158	10
XI.	159	157	155	152	149	143	149	152	155	158	159	160	161	163	164	165	164	163	161	I
10	163	161	159	156	153	150	154	157	159	162	162	163	164	166	166	168	166	166	164	50
20	166	164	163	161	158	155	159	161	163	165	166	166	168	169	169	170	169	168	167	40
30	170	168	167	165	164	161	164	166	167	169	169	170	171	172	172	173	172	171	170	30
40	173	172	171	170	169	167	169	170	171	173	173	173	174	174	175	175	175	174	174	20
50	177	176	176	175	174	174	175	175	176	176	176	177	177	177	177	177	177	177	177	10

In North Latitude the Azimuth is reckoned from North: to East when time is A.M., to West when time is P.M.
In South Latitude the Azimuth is reckoned from South: to East when time is A.M., to West when time is P.M.

(Courtesy of A. McNevin)

Fig. 78. True bearings of the sun, to be used in setting the shadow pelorus (Fig. 40) for obtaining compass deviation. The shadow will indicate true directions. For magnetic directions add westerly variation to the above values in the morning and subtract in the afternoon; easterly variation, subtract in the morning and add in the afternoon. Use sun time, not clock time. (See Fig. 77)

E S E. If the compass bearing from the boat to *A* (as indicated by the compass on the boat) is S E by E, then the deviation is 1 point westerly. As the lubberline is indicating N by E (I, Fig. 79) this, corrected for 1 point westerly deviation, shows the boat to be heading N (Magnetic).

The boat is then swung around a little (II, Fig. 79) and another compass bearing taken on *A*. Perhaps this bearing will be S E by E ½ E and as the magnetic bearing (E S E) has not changed then on this heading (N W) the compass will have a deviation of ½ point west. Swing the boat to another position (III, Fig. 79) and take a bearing on *A*. Perhaps this bearing will be E S E, which is the same as the magnetic bearing. Therefore on this heading (S W by W) the compass shows zero deviation. Proceed in the same way until the boat has been swung all around (360°) obtaining as many bearings on *A* as is convenient.

Reciprocal Bearings

Another method of determining compass deviation is by taking what are known as reciprocal bearings. This method calls for work of a more exacting nature than any of the foregoing methods, and while it is extremely serviceable for obtaining compass deviations on large boats and vessels, it is not altogether suitable for obtaining compass deviations for small boat use where it is imposible to obtain a compass bearing of an object on shore quickly, with accuracy. This method calls for placing an observer on shore, with a compass, and another observer on board the boat at the boat's compass.

A system of prearranged signals must be worked out, so that at the instant a signal is given from the boat or by the observer on shore, such as the waving of a handkerchief or the dropping of a flag, a bearing will be taken from the boat to the observer's position on shore, and at the same instant a bearing will be taken by the observer on shore of the boat's position. The bearing by the shore compass, reversed, is the magnetic bearing of the shore station from the boat. The difference between this bearing and the bearing by the boat's compass represents the deviation.

A number of such observations should be taken from the boat to the shore, and the shore to the boat, with the boat on different headings. In this way the deviation of the compass on the various magnetic headings can be determined.

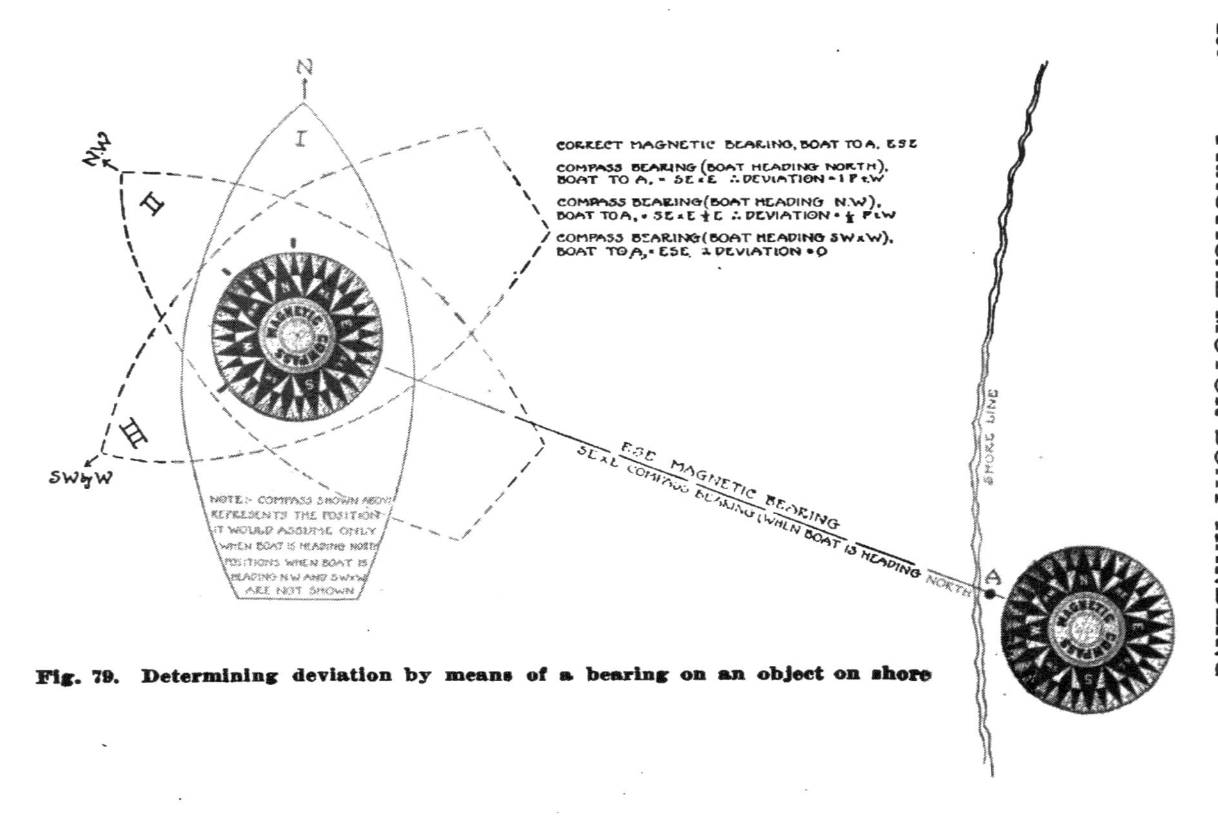

Fig. 79. Determining deviation by means of a bearing on an object on shore

CHAPTER XXIII

Compass Points and Degrees

Point	°	′	″
NORTH	0		
	0	42	11
N. ⅛ E.	1	24	
	2	06	34
N. ¼ E.	2	49	
	3	30	56
N. ⅜ E.	4	13	
	4	55	19
N. ½ E.	5	37	30
	6	19	41
N. ⅝ E.	7	02	
	7	44	04
N. ¾ E.	8	26	
	9	08	26
N. ⅞ E.	9	51	
	10	32	49
N. by E.	11	15	
	11	57	11
N. by E. ⅛ E.	12	39	
	13	21	34
N. by E. ¼ E.	14	04	
	14	45	56
N. by E. ⅜ E.	15	28	
	16	10	19
N. by E. ½ E.	16	52	30
	17	34	41
N. by E. ⅝ E.	18	17	
	18	59	04
N. by E. ¾ E.	19	41	
	20	23	26
N. by E. ⅞ E.	21	06	
	21	47	49

Point	°	′	″
NNE.	22	30	
	23	12	11
NNE. ⅛ E.	23	54	
	24	36	34
NNE. ¼ E.	25	19	
	26	00	56
NNE. ⅜ E.	26	43	
	27	25	19
NNE. ½ E.	28	07	30
	28	49	41
NNE. ⅝ E.	29	32	
	30	14	04
NNE. ¾ E.	30	56	
	31	38	26
NNE. ⅞ E.	32	21	
	33	02	49
NE. by N.	23	45	
	34	27	11
NE. ⅞ N.	35	09	
	35	51	34
NE. ¾ N.	36	34	
	37	15	56
NE. ⅝ N.	37	58	
	38	40	19
NE. ½ N.	39	22	30
	40	04	41
NE. ⅜ N.	40	47	
	41	29	04
NE. ¼ N.	42	11	
	42	53	26
NE. ⅛ N.	43	36	
	44	17	49

	°	′	″
NE.	45		
	45	42	11
NE. ⅛ E.	46	24	
	47	06	34
NE. ¼ E.	47	49	
	48	30	56
NE. ⅜ E.	49	13	
	49	55	19
NE. ½ E.	50	37	30
	51	19	41
NE. ⅝ E.	52	02	
	52	44	04
NE. ¾ E.	53	26	
	54	08	26
NE. ⅞ E.	54	51	
	55	32	49
NE. by E.	56	15	
	56	57	11
NE. by E. ⅛ E.	57	39	
	58	21	34
NE. by E. ¼ E.	59	04	
	59	45	56
NE. by E. ⅜ E.	60	28	
	61	10	19
NE. by E. ½ E.	61	52	30
	62	34	41
NE. by E. ⅝ E.	63	17	
	63	59	04
NE. by E. ¾ E.	64	41	
	65	23	26
NE. by E. ⅞ E.	66	06	
	66	47	49

	°	′	″
ENE.	67	30	
	68	12	11
ENE. ⅛ E.	68	54	
	69	36	34
ENE. ¼ E.	70	19	
	71	00	56
ENE. ⅜ E.	71	43	
	72	25	19
ENE. ½ E.	73	07	30
	73	49	41
ENE. ⅝ E.	74	32	
	75	14	04
ENE. ¾ E.	75	56	
	76	38	26
ENE. ⅞ E.	77	21	
	78	02	49
E. by N.	78	45	
	79	27	11
E. ⅞ N.	80	09	
	80	51	34
E. ¾ N.	81	34	
	82	15	56
E. ⅝ N.	82	58	
	83	40	19
E. ½ N.	84	22	30
	85	04	41
E. ⅜ N.	85	47	
	86	29	04
E. ¼ N.	87	11	
	87	53	26
E. ⅛ N.	88	36	
	89	17	49

Point	°	′	″
EAST	90		
	90	42	11
E. ⅛ S.	91	24	
	92	06	34
E. ¼ S.	92	49	
	93	30	56
E. ⅜ S.	94	13	
	94	55	19
E. ½ S.	95	37	30
	96	19	41
E. ⅝ S.	97	02	
	97	44	04
E. ¾ S.	98	26	
	99	08	26
E. ⅞ S.	99	51	
	100	32	49
E. by S.	101	15	
	101	57	11
ESE. ⅞ E.	102	39	
	103	21	34
ESE. ¾ E.	104	04	
	104	45	56
ESE. ⅝ E.	105	28	
	106	10	19
ESE. ½ E.	106	52	30
	107	34	41
ESE. ⅜ E.	108	17	
	108	59	04
ESE. ¼ E.	109	41	
	110	23	26
ESE. ⅛ E.	111	06	
	111	47	49
ESE	112	30	
	113	12	11
SE. by E. ⅞ E.	113	54	
	114	36	34
SE. by E. ¾ E.	115	19	
	116	00	56
SE. by E. ⅝ E.	116	43	
	117	25	19
SE. by E. ½ E.	118	07	30
	118	49	41
SE. by E. ⅜ E.	119	32	
	120	14	04
SE. by E. ¼ E.	120	56	
	121	38	26
SE. by E. ⅛ E.	122	21	
	123	02	49
SE. by E.	123	45	
	124	27	11
SE. ⅞ E.	125	09	
	125	51	34
SE. ¾ E.	126	34	
	127	15	56
SE. ⅝ E.	127	58	
	128	40	19
SE. ½ E.	129	22	30
	130	04	41
SE. ⅜ E.	130	47	
	131	29	04
SE. ¼ E.	132	11	
	132	53	26
SE. ⅛ E.	133	36	
	134	17	49

	°	′	″
SE.	135		
	135	42	11
SE. ⅛ S.	136	24	
	137	06	34
SE. ¼ S.	137	49	
	138	30	56
SE. ⅜ S.	139	13	
	139	55	19
SE. ½ S.	140	37	30
	141	19	41
SE. ⅝ S.	142	02	
	142	44	04
SE. ¾ S.	143	26	
	144	08	26
SE. ⅞ S.	144	51	
	145	32	49
SE. by S.	146	15	
	146	57	11
SSE. ⅞ E.	147	39	
	148	21	34
SSE. ¾ E.	149	04	
	149	45	56
SSE. ⅝ E.	150	28	
	151	10	19
SSE. ½ E.	151	52	30
	152	34	41
SSE. ⅜ E.	153	17	
	153	59	04
SSE. ¼ E.	154	41	
	155	23	26
SSE. ⅛ E.	156	06	
	156	47	49
SSE.	157	30	
	158	12	11
S. by E. ⅞ E.	158	54	
	159	36	34
S. by E. ¾ E.	160	19	
	161	00	56
S. by E. ⅝ E.	161	43	
	162	25	19
S. by E. ½ E.	163	07	30
	163	49	41
S. by E. ⅜ E.	164	32	
	165	14	04
S. by E. ¼ E.	165	56	
	166	38	26
S. by E. ⅛ E.	167	21	
	168	02	49
S. by E.	168	45	
	169	27	11
S. ⅞ E.	170	09	
	170	51	34
S. ¾ E.	171	34	
	172	15	56
S. ⅝ E.	172	58	
	173	40	19
S. ½ E.	174	22	30
	175	04	41
S. ⅜ E.	175	47	
	176	29	04
S. ¼ E.	177	11	
	177	53	26
S. ⅛ E.	178	36	
	179	17	49

Point	°	′	″
SOUTH	180		
	180	42	11
S. ⅛ W.	181	24	
	182	06	34
S. ¼ W.	182	49	
	183	30	56
S. ⅜ W.	184	13	
	184	55	19
S. ½ W.	185	37	30
	186	19	41
S. ⅝ W.	187	02	
	187	44	04
S. ¾ W.	188	26	
	189	08	26
S. ⅞ W.	189	51	
	190	32	49
S. by W.	191	15	
	191	57	11
S. by W. ⅛ W.	192	39	
	193	21	34
S. by W. ¼ W.	194	04	
	194	45	56
S. by W. ⅜ W.	195	28	
	196	10	19
S. by W. ½ W.	196	52	30
	197	34	41
S. by W. ⅝ W.	198	17	
	198	59	04
S. by W. ¾ W.	199	41	
	200	23	26
S. by W. ⅞ W.	201	06	
	201	47	49

Point	°	′	″
SSW	202	30	
	203	12	11
SSW. ⅛ W.	203	54	
	204	36	34
SSW. ¼ W.	205	19	
	206	00	56
SSW. ⅜ W.	206	43	
	207	25	19
SSW. ½ W.	208	07	30
	208	49	41
SSW. ⅝ W.	209	32	
	210	14	04
SSW. ¾ W.	210	56	
	211	38	26
SSW. ⅞ W.	212	21	
	213	02	49
SW. by S.	213	45	
	214	27	11
SW. ⅞ S.	215	09	
	215	51	34
SW. ¾ S.	216	34	
	217	15	56
SW. ⅝ S.	217	58	
	218	40	19
SW. ½ S.	219	22	30
	220	04	41
SW. ⅜ S.	220	47	
	221	29	04
SW. ¼ S.	222	11	
	222	53	26
SW. ⅛ S.	223	36	
	224	17	49

	°	′	″
SW.	225		
	225	42	11
SW. ⅛ W.	226	24	
	227	06	34
SW. ¼ W.	227	49	
	228	30	56
SW. ⅜ W.	229	13	
	229	55	19
SW. ½ W.	230	37	30
	231	19	41
SW. ⅝ W.	232	02	
	232	44	04
SW. ¾ W.	233	26	
	234	08	26
SW. ⅞ W.	234	51	
	235	32	49
SW. by W.	236	15	
	236	57	11
SW. by W. ⅛ W.	237	39	
	238	21	34
SW. by W. ¼ W.	239	04	
	239	45	56
SW. by W. ⅜ W.	240	28	
	241	10	19
SW. by W. ½ W.	241	52	30
	242	34	41
SW. by W. ⅝ W.	243	17	
	243	59	04
SW. by W. ¾ W.	244	41	
	245	23	26
SW. by W. ⅞ W.	246	06	
	246	47	49
WSW.	247	30	
	248	12	11
WSW. ⅛ W.	248	54	
	249	36	34
WSW. ¼ W.	250	19	
	251	00	56
WSW. ⅜ W.	251	43	
	252	25	19
WSW. ½ W.	253	07	30
	253	49	41
WSW. ⅝ W.	254	32	
	255	14	04
WSW. ¾ W.	255	56	
	256	38	26
WSW. ⅞ W.	257	21	
	258	02	49
W. by S.	258	45	
	259	27	11
W. ⅞ S.	260	09	
	260	51	34
W. ¾ S.	261	34	
	262	15	56
W. ⅝ S.	262	58	
	263	40	19
W. ½ S.	264	22	30
	265	04	41
W. ⅜ S.	265	47	
	266	29	04
W. ¼ S.	267	11	
	267	53	26
W. ⅛ S.	268	36	
	269	17	49

	°	′	″
WEST	270		
	270	42	11
W. ⅛ N.	271	24	
	272	06	34
W. ¼ N.	272	49	
	273	30	56
W ⅜ N.	274	13	
	274	55	19
W. ½ N.	275	37	30
	276	19	41
W. ⅝ N.	277	02	
	277	44	04
W. ¾ N.	278	26	
	279	08	26
W. ⅞ N.	279	51	
	280	32	49
W. by N.	281	15	
	281	57	11
WNW. ⅞ W.	282	39	
	283	21	34
WNW. ¾ W.	284	04	
	284	45	56
WNW. ⅝ W.	285	28	
	286	10	19
WNW. ½ W.	286	52	30
	287	34	41
WNW. ⅜ W.	288	17	
	288	59	04
WNW ¼ W.	289	41	
	290	23	26
WNW. ⅛ W.	291	06	
	291	47	49
WNW.	292	30	
	293	12	11
NW. by W. ⅞ W.	293	54	
	294	36	34
NW. by W. ¾ W.	295	19	
	296	00	56
NW. by W. ⅝ W.	296	43	
	297	25	19
NW. by W. ½ W.	298	07	30
	298	49	41
NW. by W. ⅜ W.	299	32	
	300	14	04
NW. by W ¼ W.	300	56	
	301	38	26
NW. by W. ⅛ W.	302	21	
	303	02	49
NW. by W.	303	45	
	304	27	11
NW. ⅞ W.	305	09	
	305	51	34
NW. ¾ W.	306	34	
	307	15	56
NW. ⅝ W.	307	58	
	308	40	19
NW. ½ W.	309	22	30
	310	04	41
NW. ⅜ W.	310	47	
	311	29	04
NW. ¼ W.	312	11	
	312	53	26
NW. ⅛ W.	313	36	
	314	17	49

Point	°	′	″
NW.	315		
	315	42	11
NW. ⅛ N.	316	24	
	317	06	34
NW. ¼ N.	317	49	
	318	30	56
NW. ⅜ N.	319	13	
	319	55	19
NW. ½ N.	320	37	30
	321	19	41
NW. ⅝ N.	322	02	
	322	44	04
NW. ¾ N.	323	26	
	324	08	26
NW. ⅞ N.	324	51	
	325	32	49
NW. by N.	326	15	
	326	57	11
NNW. ⅞ W.	327	39	
	328	21	34
NNW. ¾ W.	329	04	
	329	45	56
NNW. ⅝ W.	330	28	
	331	10	19
NNW. ½ W.	331	52	30
	332	34	41
NNW. ⅜ W.	333	17	
	333	59	04
NNW. ¼ W.	334	41	
	335	23	26
NNW. ⅛ W.	336	06	
	336	47	49
NNW.	337	30	
	338	12	11
N. by W. ⅞ W.	338	54	
	339	36	34
N. by W. ¾ W.	340	19	
	341	00	56
N. by W. ⅝ W.	341	43	
	342	25	19
N. by W. ½ W.	343	07	30
	343	49	41
N. by W. ⅜ W.	344	32	
	345	14	04
N. by W. ¼ W.	345	56	
	346	38	26
N. by W. ⅛ W.	347	21	
	348	02	49
N. by W.	348	45	
	349	27	11
N. ⅞ W.	350	09	
	350	51	34
N. ¾ W.	351	34	
	352	15	56
N. ⅝ W.	352	58	
	353	40	19
N. ½ W.	354	22	30
	355	04	41
N. ⅜ W.	355	47	
	356	29	04
N. ¼ W.	357	11	
	357	53	26
N. ⅛ W.	358	36	
	359	17	49

Lights for Various Types of Boats

(See Chapter III).

	Bow	Side	Stern	Masthead	Special
1 Motor boat, Class 1	R*G*		(W)		
2 Motor boat, Classes 2 and 3	W**	R*G*	(W)		
3 Motor boat (under sail and power)		R*G*			(W)[6]
4 Inland steamer	W**	R*G*	(W)		
5 Ocean-going steamer	W**(1)	R*G*	W*		W**[2]
6 Ocean-going yacht	W**(1)	R*G*			W**[2]
7 Sailing vessel		R*G*	(W)[6]		
8 Harbor tug (towing one boat)		R*G*		(W) (W)	
9 Harbor tug (towing two or more boats astern)[7]		R*G*		(W) (W) (W)	
10 Dredges (stationary)				(R) (R)	(W)(W)(W)(W)[5]
11 Ferry boat		R*G*			(W)(W)[9]
12 Ocean-going tug (towing one boat)		R*G*		W** W**	
13 Ocean-going tug (towing two or more boats astern)[8]		R*G*	W*	W** W**	
14. Ocean-going barges being towed		R*G*	W*		(W)(W)[3]
15 Inland barges, canal boats, etc. (tandem)	(W)		(W)		(W)(W)[3]
16 Pilot vessel (under way)		R*G*		(W) (R)	
17. Pilot vessel (on station at anchor)				(W) (R)	
18. Fishing vessel		R*G*		(R) (W)	
19. Boat at anchor (less than 150 feet)	(W)				
20. Boat at anchor (more than 150 feet)	(W)		(W)		
21. Government patrol vessel		R*G*		(W) (R) (W)	
22 Vessel under partial control only		R*G*		(R) (R)	
23. Vessel totally disabled				(R) (R)	

Lights for Various Types of Boats—Continued

	Bow	Side	Stern	Masthead	Special
24. Vessel laying cable, etc.		R*G*		(R) (W) (R)	
25. Vessel aground in channel	(W)			(R) (R)	
26. Vessel towing a wreck.		R*G*		(W) (R) (R) (W)	
27. Vessel made fast to or moored over a wreck				(R) (R)	(W)[4]

Key

* = Light showing for 10 points
** = " " " 20 " ahead
( ) = " " " 36 "
a = " " " 12 " astern
W = White light
R = Red (port light)
G = Green (starboard)
(1) Placed on foremast
(2) Placed on mainmast
(3) On stern of last barge in tow only
(4) White light displayed at bow and stern of each outside vessel or lighter
(5) White light displayed at each corner
(6) White light displayed upon being overtaken by another boat
(7) When length of tow exceeds 600 feet; otherwise lighted as per No. 8
(8) When length of tow exceeds 600 feet otherwise lighted as per No. 12
(9) Placed at equal altitudes above the water, bow and stern

CHAPTER XXIV

Navigation Wrinkles

In piloting and sailing along the coast several methods were mentioned in Chapter XI for locating one's position. The commoner methods include, besides cross bearings, several schemes which take two bearings on a single object, the bearings being separated by a distance run by the boat, such as doubling the angle, bow and beam bearing or 45-90-degree method, etc. Most of these permit of being worked out graphically on the chart as well as mathematically.

For example, in Fig. 53 (page 86) at some place *O*, take a bearing of the light *A* and draw a line on the chart representing the direction of the bearing just taken. At the time this bearing is taken note the time or the reading of the log. Hold the boat on her course until the bearing of the light has changed at least 30 degrees, and then take another bearing, noting the time *D* or the log reading. Transfer this bearing to the chart and draw a line from *A* representing this bearing. Mark off on another piece of paper the distance traveled between bearings and keeping this paper parallel with the direction of the course, move it in toward *A*. At a certain place one end of the paper will coincide with the line *OA* and the other with *OP*. This will locate point *P* and will be the boat's position at the time of taking the second bearing.

Doubling the angle (described on page 87) is a well-known method of locating one's position. The two bearings may be either in points or degrees, that is two points-four points; three points-six points; 45-90 degrees; 30-60 degrees, etc., and the result will be exactly the same, i. e., the distance off the object at the time the second bearing is taken will be equal to the distance run between bearings.

By taking a bearing on an object when it is abeam, together with another bearing of the same object either before or after it is abeam one will obtain a series of right angle triangles which make many problems of location very easily solved. For example (see Fig. 81), if any *one* (or more) of the distances *a, b, c, d, b', c', d', AB, AC, AD,* etc., are known (these may be known very easily by obvious methods) then any of the other distances can be readily computed from the table on page 162. Thus the boat's location can be immediately determined.

SHORT CUTS IN POSITION LOCATING

(See Fig. 81)

a = 0.9b = 0.7c = 0.5d = 1.7AB = 2.4BC = 1.4CD = 3.6DE = 1.0AC = 0.6AD = 0.5AE = 0.9BD = 0.7BE

b = b' = 1.2a = 0.8c = 0.6d = 2.0AB = 2:8BC = 1.6CD = 4.1DE = 1.2AC = 0.7AD = 0.6AE = 1.0BD = 0.8BE

c = c' = 1.4a = 1.2b = 0.7d = 2.4AB = 3.3BC = 2.0CD = 5.0DE = 1.4AC = 0.8AD = 0.7AE = 1.2BD = 1.0BE

d = d' = 2.0a = 1.8b = 1.4c = 3.5AB = 4.8BC = 2.8CD = 7.1DE = 2.0AC = 1.2AD = 1.0AE = 1.8BD = 1.4BE

AB = AB' = 0.6a = 0.5b = 0.4c = 0.3d = 1.4BC = 0.8CD = 2.0DE = 0.6AC = 3.3AD = 0.3AE = 0.5BD = 0.4BE

BC = B'C' = 0.4a = 0.4b = 0.3c = 0.2d = 0.7AB = 0.6CD = 1.5DE = 0.4AC = 0.2AD = 0.2AE = 0.4BD = 0.3BE

CD = C'D' = 0.7a = 0.6b = 0.5c = 0.4d = 1.2AB = 1.7BC = 2.5DE = 0.7AC = 0.4AD = 0.4AE = 0.6BD = 0.5BE

DE = 0.3a = 0.2b = 0.2c = 0.1d = 0.5AB = 0.7BC = 0.4CD = 0.3AC = 0.2AD = 0.1AE = 0.2BD = 0.2BE

AC = AC' = 1.0a = 0.9b = 0.7c = 0.5d = 1.7AB = 2.4BC = 1.4CD = 3.6DE = 0.6AD = 0.5AE = 0.9BD = 0.7BE

AD = AD' = 1.7a = 1.5b = 1.2c = 0.9d = 3.0AB = 4.1BC = 2.4CD = 6.3DE = 1.7AC = 0.9AE = 1.5BD = 1.2BE

AE = 2.0a = 1.7b = 1.4c = 1.0d = 3.5AB = 4.8BC = 2.8CD = 7.1DE = 2.0AC = 1.2AD = 1.8BD = 1.4BE

BD = B'D' = 1.2a = 1.0b = 0.8c = 0.6d = 2.0AB = 2.7BC = 1.6CD = 4.1DE = 1.2AC = 0.7AD = 0.6AE = 0.8BE

BE = 1.4a = 1.2b = 1.0c = 0.7d = 2.4AB = 3.3BC = 2.0CD = 5.0DE = 1.4AC = 0.8AD = 0.7AE = 1.2BD

Examples: (See Fig. 81.) Distance *a* is found to be 4 miles, how far from *O* will the position of the boat be when *O* is 30° abaft the beam? Referring to the above table it will be seen that *b* = 1.2a, therefore, point *B'* is located 4.8 (1.2 X 4) miles from *O*.

A boat has sailed 2.5 miles between a 30° bearing (*D*) and a 45° bearing (*C*) of a certain object. How far off the object is she: (1) when the 45° bearing is taken; (2) when the 30° was taken? From the above table, c = 2 X CD and d = 2.8 CD; therefore, at the time of taking the 45° bearing the boat must have been 5 (2 X 2.5) miles away from the object. At the time the 30° bearing was taken the boat was 7.0 (2.8 X 2.5) miles away from the object.

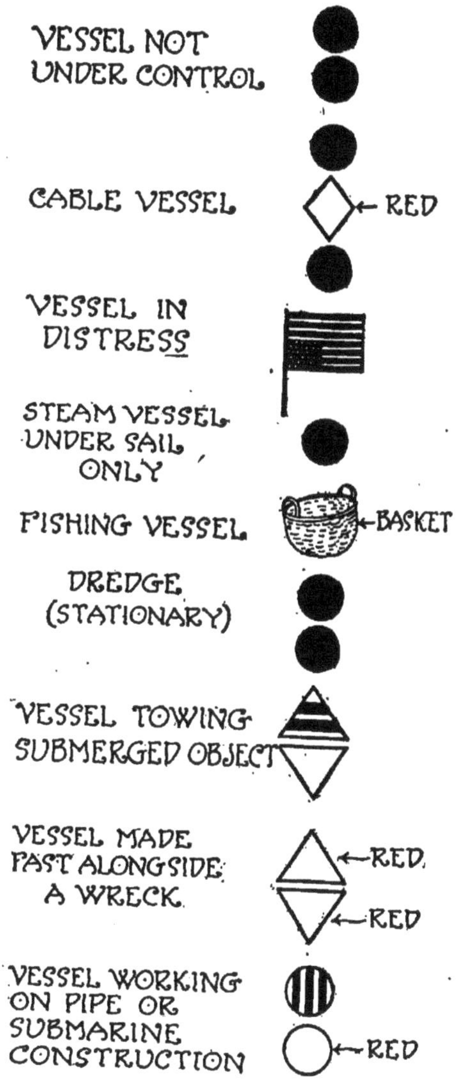

VESSEL NOT UNDER CONTROL

CABLE VESSEL ← RED

VESSEL IN DISTRESS

STEAM VESSEL UNDER SAIL ONLY

FISHING VESSEL ← BASKET

DREDGE (STATIONARY)

VESSEL TOWING SUBMERGED OBJECT

VESSEL MADE FAST ALONGSIDE A WRECK ← RED ← RED

VESSEL WORKING ON PIPE OR SUBMARINE CONSTRUCTION ← RED

Fig. 80. Day marks

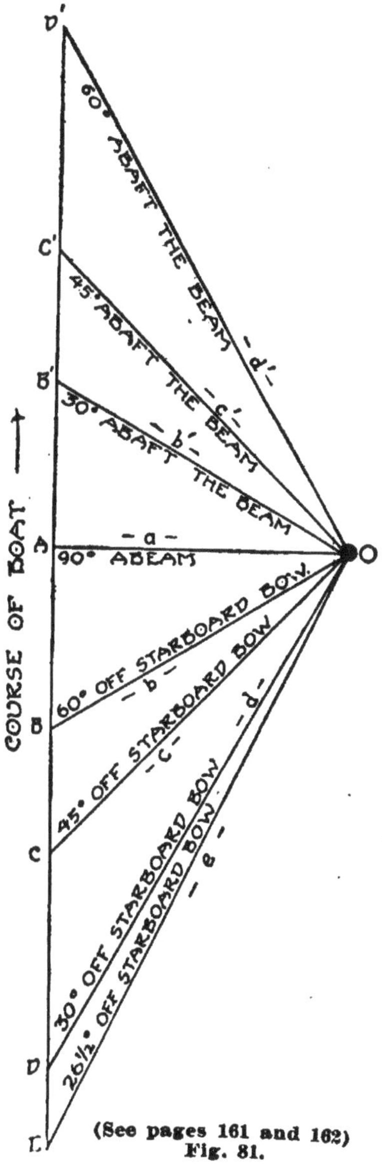

D'
60° ABAFT THE BEAM – d' –
C'
45° ABAFT THE BEAM – c' –
B'
30° ABAFT THE BEAM – b' –
A
– a –
90° ABEAM
60° OFF STARBOARD BOW – b –
B
45° OFF STARBOARD BOW – c –
C
30° OFF STARBOARD BOW – d –
D
26½° OFF STARBOARD BOW – e –
E
COURSE OF BOAT
O

(See pages 161 and 162)
Fig. 81.

CHAPTER XXV

Naval Insignia

Shoulder Straps

Upper row: Rear Admiral, Lieutenant-Commander, Paymaster, and Ensign. Lower row: Chief Gunner, Gunner, Chief Boatswain, and Boatswain

Above: Carpenter, Warrant Machinist, Mate, and Pay Clerk

Metal Insignia: Rear Admiral, Commander, and Ensign

Collar Devices

Admiral of the Navy

Admiral (should Congress revive the grade)

Vice-Admiral (should Congress revive the grade)

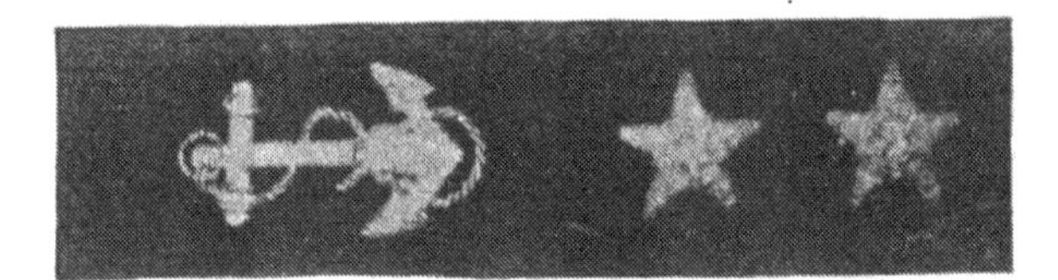

Rear Admiral

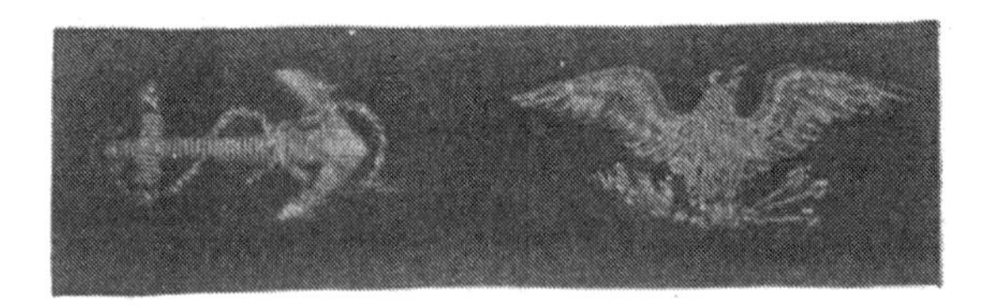

Captain

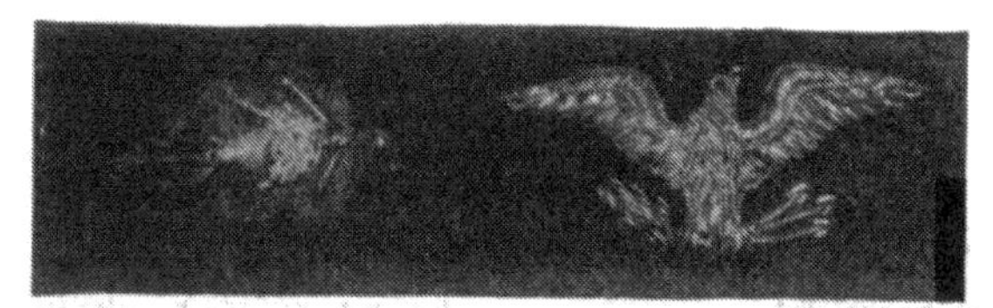

Medical Officer, rank of Captain

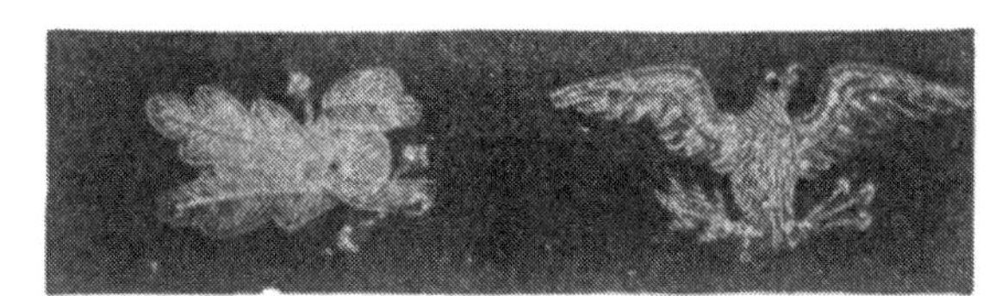

Pay Officer, rank of Captain

Chaplain, rank of Commander

Naval Constructor, rank of Commander

Civil Engineer, rank of Commander

Professor of Mathematics, rank of Commander

Dental Officer, rank of Lieutenant

Collar Devices on Service Coat

Left hand column, reading downward: Sailmaker, Pharmacist, Paymaster's Clerk, Mate. Right hand column, reading downward: Chief Boatswain, Chief Gunner, Chief Machinist, Chief Carpenter

Insignia of Rank on Sleeve

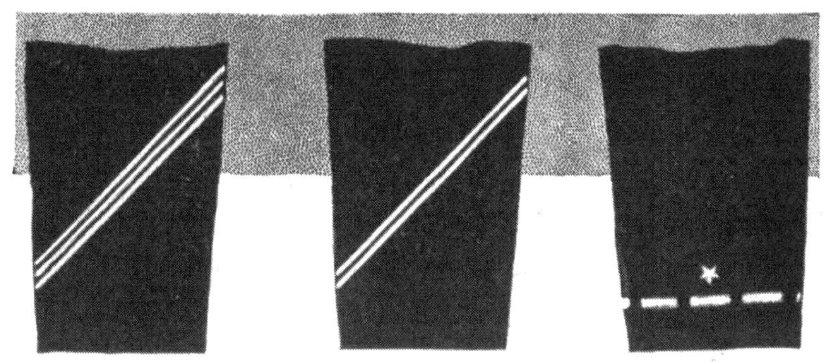

Midshipman first class, Midshipman, second class, Chief Boatswain
Chief Gunner
Chief Machinist

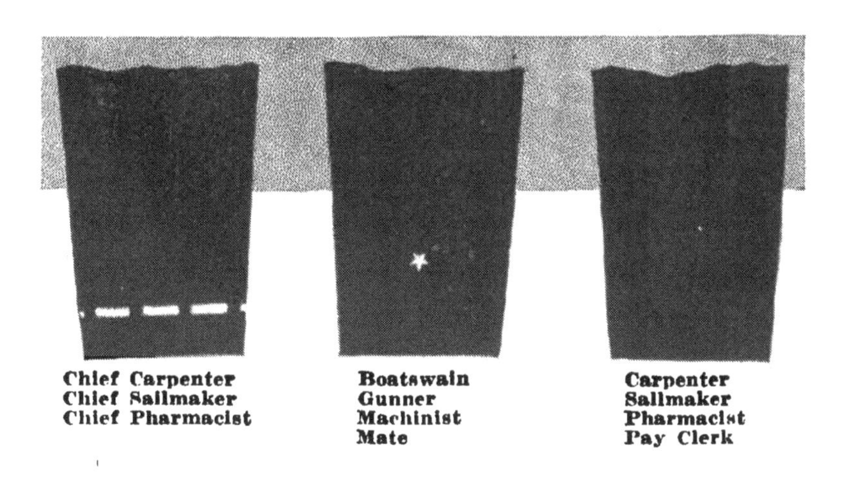

Chief Carpenter
Chief Sailmaker
Chief Pharmacist

Boatswain
Gunner
Machinist
Mate

Carpenter
Sailmaker
Pharmacist
Pay Clerk

Admiral of the Navy, Admiral, Vice-Admiral, Rear Admiral, and Captain

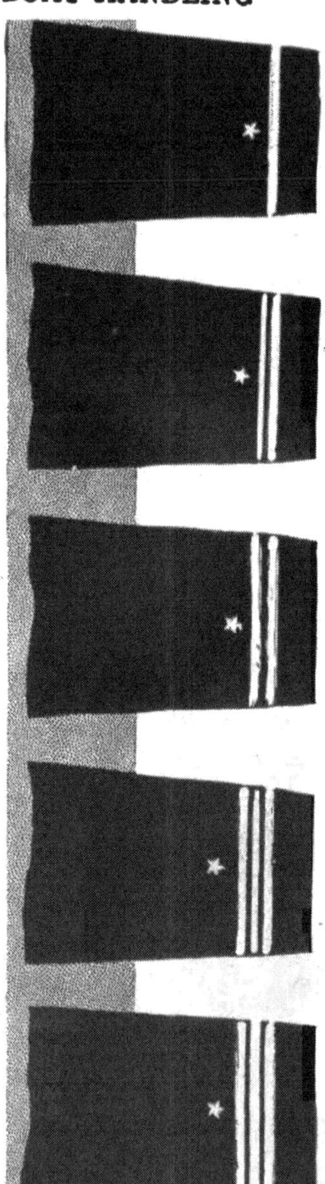

Commander, Lieutenant-Commander, Lieutenant, Lieutenant (J. G.), and Ensign

Rating Badges (Blue)

Chief Master-at-Arms

Boatswain's Mate
(First Class)

Gunner's Mate
(Second Class)

Quartermaster
(Third Class)

INDEX

Milton Keynes UK
Ingram Content Group UK Ltd.
UKHW022022170823
427026UK00007B/359